CLOSE UP AT A DISTANCE

CLOSE UP AT A DISTANCE

MAPPING, TECHNOLOGY, AND POLITICS

Laura Kurgan

ZONE BOOKS · NEW YORK · 2013

Printed in Canada.

Distributed by The MIT Press,
Cambridge, Massachusetts, and London, England

Library of Congress Cataloging-in-Publication Data

Kurgan, Laura.
        Close up at a distance : mapping, technology, and politics
/ Laura Kurgan.
        p.   cm.
    ISBN 978-1-935408-28-4 (alk. paper)
        1. Satellite image maps. 2. Remote-sensing images—
Political aspects. 3. Global positioning system—Social aspects.
4. Aerial photography. I. Title.

G70.4.K87 2013
526—dc23
                                2012040173

Contents

INTRODUCTION

Astronaut photograph AS8-14-2383, December 24, 1968. NASA's original caption reads: "This view of the rising Earth greeted the Apollo 8 astronauts as they came from behind the Moon after the lunar orbit insertion burn. Earth is about five degrees above the horizon in the photo. The unnamed surface features in the foreground are near the eastern limb of the Moon as viewed from Earth. The lunar horizon is approximately 780 kilometers from the spacecraft. Width of the photographed area at the horizon is about 175 kilometers. On the Earth 240,000 miles away, the sunset terminator bisects Africa." This image has come to be known as *Earthrise*. PHOTO: NASA

Astronaut photograph AS17-148-22727, December 7, 1972. NASA's original caption reads: "View of the Earth as seen by the Apollo 17 crew traveling toward the moon. This translunar coast photograph extends from the Mediterranean Sea area to the Antarctica south polar ice cap. This is the first time the Apollo trajectory made it possible to photograph the south polar ice cap. Note the heavy cloud cover in the Southern Hemisphere. Almost the entire coastline of Africa is clearly visible. The Arabian Peninsula can be seen at the northeastern edge of Africa. The large island off the coast of Africa is the Malagasy Republic. The Asian mainland is on the horizon toward the northeast." This image has come to be known as *The Blue Marble*. PHOTO: NASA

Mapping Considered as a
Problem of Theory and Practice

Consider two similar images that have transcended mere publicity to become iconic. *Earthrise*, or image AS8-14-2383, is a color photograph taken by Apollo 8 astronaut William Anders in December 1968, showing the Earth in half shadow against the foreground of a lunar landscape. The second picture comes from the Apollo 17 astronauts in December 1972, a circular image of a shadowless globe. NASA labeled it image number AS17-148-22727, but it has come to be called *The Blue Marble*.

Earthrise is a photo of the Earth taken while orbiting the Moon. It is a perspectival view — the foreground offers a sort of ground and seems to suggest the position of a viewer, so that you can almost imagine being there, looking across the lunar surface. *The Blue Marble* is perhaps more unsettling, because it is without perspective, a floating globe, an abstracted sphere, something like a map.

Denis Cosgrove, in *Apollo's Eye*, calls our attention to these two images and to the role they played in producing "an altered image of the Earth."[1] Each in its own way is credited with representing or even catalyzing a notion of global or planetary unity, whether in universalist terms, humanist ones, or precisely non-humanist environmental or natural ones. The view across the Moon's surface, it seems, provoked thoughts of an Earth without borders. Cosgrove quotes Apollo 8 mission commander Frank Borman's reading of the *Earthrise* image: "When you're finally up at the moon looking back at the earth, all those differences and nationalistic traits are pretty well going to blend and you're going to get a concept that maybe this is really one world and why the hell can't we learn to live together like decent people?"[2] This "concept" of "one world" can be evaluated in many ways: as "the universal brotherhood of a common humanity" (Cosgrove paraphrasing Archibald MacLeish), as a gesture of imperial domination, as an abstract and artificially totalizing erasure of very real differences, as the basis of new global political

The *Blue Marble 2002* is a composite image stitching together quarterly observations, at a spatial resolution of 1 square kilometer per pixel, from the Moderate Resolution Imaging Spectroradiometer (MODIS) onboard NASA's Terra satellite.

NASA IMAGE BY RETO STÖCKLI WITH ENHANCEMENTS BY ROBERT SIMMON; ADDITIONAL DATA FROM USGS EROS DATA CENTER, USGS TERRESTRIAL REMOTE SENSING FLAGSTAFF FIELD CENTER (ANTARCTICA), AND DEFENSE METEOROLOGICAL SATELLITE PROGRAM

movements for human rights or planetary responsibility, or as what Martin Heidegger called "the uprooting of man"—"I was shocked when a short time ago I saw the pictures of the earth taken from the moon. We do not need atomic bombs at all—the uprooting of man is already here…. It is no longer upon an earth that man lives today," he told an interviewer in 1966, just a month after an even earlier *Earthrise* image, taken from the Lunar Orbiter 1, had been released.[3] Whatever the evaluation, as Cosgrove underlines, these photographs "have become the image of the globe, simultaneously 'true' representations and virtual spaces."[4] The 1972 photograph, no doubt because it both offered the viewer the whole Earth and seemed to remove any viewer from the picture, became perhaps even more of an icon, not only of totality and unity but likewise singularity and freestanding vulnerability.

But these two images are not the only examples of their type, and their afterlife is indicative of an important shift in the way we represent the planet—and the political stakes of those representations. The iconic status of the images, particularly the second one, is perhaps attested to by the fact that most people will not be able to notice a difference between the 1972 *Blue Marble* and a number of new ones. In 2002, NASA produced a pair of new images, together called once again

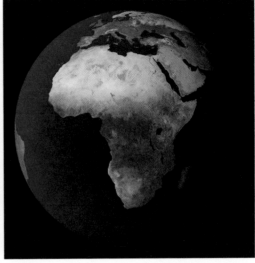

The Blue Marble: The Next Generation, 2005, is a composite image using twelve monthly cloud-free observations in 2004, at a spatial resolution of 500 square meters per pixel, from the MODIS onboard NASA's Terra satellite.   IMAGE: RETO STÖCKLI, NASA EARTH OBSERVATORY

The Blue Marble (one of the Western Hemisphere, and one of the Eastern), put together out of four months' worth of satellite images assembled into what the space agency called a "seamless, photo-like mosaic of every square kilometer of our planet." The resolution of the images, collected by the Moderate Resolution Imaging Spectroradiometer, was one kilometer per pixel. Three years later, they did it again, at twice the resolution and based on twelve months' worth of remote sensing, and called the images *The Blue Marble: The Next Generation*.[5] And in 2012, there were two more, again one of the Western Hemisphere and the other of the Eastern, called *Blue Marble Next Generation 2012*, assembled from data collected by the Visible/Infrared Imager Radiometer Suite (VIIRS) on the Suomi NPP satellite in six orbits over eight hours.[6] These versions are not simply photographs taken by a person traveling in space with a camera. They are composites of massive quantities of remotely sensed data collected by satellite-borne sensors.[7]

The difference between the generations of *Blue Marbles* sums up a shift in ways of thinking about images, what they represent, and how we are to interpret them.

The new blue marbles now appear everywhere: in advertisements and as the ubiquitous default screen of the iPhone.[8] So where you might think you're looking at image number AS17-148-22727, handcrafted witness to earthly totality, in fact what you're seeing is a patchwork of satellite data, artificially assembled—albeit

Blue Marble Next Generation, 2012, is a composite image using a number of swaths of the Earth's surface taken on January 4, 2012, by the VIIRS instrument aboard NASA's Suomi NPP satellite. IMAGE: NASA/NOAA/GSFC/SUOMI NPP/VIIRS/NORMAN KURING

with great skill and an enormous amount of labor. This is not the integrating vision of a particular person standing in a particular place or even floating in space. It's an image of something no human could see with his or her own eye, not only because it's cloudless, but because it's a full 360-degree composite, made of data collected and assembled over time, wrapped around a wireframe sphere to produce a view of the Earth at a resolution of at least half a kilometer per pixel—and any continent can be chosen to be in the center of the image. As the story of the versions suggests, it can always be updated with new data. It bears with it a history that mixes, unstably, both precision and ambiguity and that raises a series of fundamental questions about the intersection between physical space and its representation, virtual space and its realization.

Cosgrove described the astronauts' photographs as "simultaneously 'true' representations and virtual spaces," and we can now begin to appreciate just how precise that description is for the sequence of satellite-generated images to which they gave rise. The photographs were true, at least in the trivial mechanical sense, and then provided a platform for something more abstract or virtual, the "concept" of "one world." Now it is the virtuality of the data-based constructions that seems self-evident. And their basis in remotely sensed data helps us understand what has become of truth in the era of the digital data stream: it is intimately related to resolution, to measurability, to the construction of a reliable algorithm for translating between representation and reality. The fact that they are virtual images does not make them any less true, but it should make us pause and consider what we mean today by truth.

It is the intersection between the true and the virtual that is the subject of what follows. In it, I offer an account of the technologies that produce global imagery and that both necessitate and facilitate the interpretation of images at once measurable and digital, uncentered and ambiguous, yet comprehensive and authoritative. My account rests on and results from research conducted through practice, working with maps and images I have created, data I have acquired or generated, installations and projects I have undertaken.

RESEARCH CONDUCTED THROUGH PRACTICE

Since the early 1990s—since the first Gulf War, to be precise—I have been thinking about and working with new technologies of location, remote sensing, and mapping. I understand this work as a form of research conducted through practice. The propositions and claims I offer here, however theoretical they are, only emerged for me through the process of experimenting with the technologies themselves, working with and through them to create images. That research has not simply been aimed at developing a theoretical framework for better understanding these new sorts of spatial representations, but has taken the form of a series of projects utilizing the technologies that have produced these images in order to investigate them. That work is presented here in terms of a series of projects that have formed the basis of my inquiry. They both exemplify the approach to understanding digital images articulated here and, I hope, suggest further lines of exploration.

The technologies of global positioning, imaging, and interpretation made available by the development of satellites tasked with surveillance and mapping first emerged to serve the needs of governments and their military and intelligence establishments. Subsequently, these technologies have been made available to the public for commercial and other ends. In the projects documented here, my aims were neither military nor commercial, but while many began as exhibitions in art galleries or museums and then were extended in print and online, they have been no less political than those of the governments and militaries that underwrote the technologies in the first place. This book gathers and reframes a number of these projects in order to make claims and arguments about what the technologies of spatial representation have to do with the spaces they represent, beyond simply representing them.

It offers a series of images created as the once-classified government and military digital technologies of mapping became publicly available, and with them the data on which they rely. In a certain sense, these images are nothing but maps, although not in the ordinary sense. Maps construct space—physical, propositional, discursive, political, archival, and memorial spaces. For many of us, maps now are as omnipresent as the more obvious utilities (such as electricity, water, gas, telephone, television, the Internet), functioning somehow like "extensions" of ourselves, to co-opt Marshall McLuhan's famous definition of media. They have become infrastructures and systems, and we are located, however insecurely, within them. Drawn with satellites, assembled with pixels radioed from outer space, and constructed out of statistics joined to specific geographies, the maps presented here record situations of intense conflict and struggle, on the one hand, and fundamental transformations in our ways of seeing and of experiencing space, on the other.

Central to the ways these projects unfolded and to the fact that they do not simply analyze, but in fact employ, these technologies, is this claim: we do not stand at a distance from these technologies, but are addressed by and embedded within them. These projects explicitly reject the ideology, the stance, and the security of "critical distance" and reflect a basic operational commitment to a practice that explores spatial data and its processing from within. Only through a certain intimacy with these technologies—an encounter with their opacities, their assumptions, their intended aims—can we begin to assess their full ethical and political stakes.[9]

These projects were made possible by and unfolded in reaction to a series of events over the last two decades that amount to a cataclysmic shift in our ability to navigate, inhabit, and define the spatial realm. They were brought on by: the operationalizing of Global Positioning System (GPS) satellites for both military and civilian uses in 1991; the democratization and distribution of data and imagery on the World Wide Web in 1992; the proliferation of desktop computing and the use of geographic information systems for the management of data; the privatization of commercial high-resolution satellites later in the 1990s; and widespread mapping made possible by Google Earth in 2005. They are also conditioned by and explore a series of political, military, and social conflicts that have defined what is loosely called the "post–Cold War" period, a time in which war fighting became ever more deeply invested in image and information technologies and in which the borders between the civilian and the military, the domestic and the international, became more and more blurred. Each project captures a moment in time politically and, with the technical means possible at that moment, zooms in and expands that moment in space and time, with all the complexities entailed in the repurposing of any image from its intended functions to new ones.

A THEORY MACHINE

Toward the end of *Einstein's Clocks, Poincaré's Maps*, Peter Galison insists on the ways in which, in the twentieth century, "machines tied clocks and maps ever closer together." He focuses on the systems constructed by "American defense planners" that "turned satellites into radio stations that would beam timed signals to earth." In that transmission, an extremely precise accounting of time can translate into an extremely accurate recording of location: "50 billionths of a second per day provide[s] a resolution on the earth's surface of fifty feet."[10]

But the accuracy is, Galison argues, *relative* — indeed, the entire operation is for him a sort of concrete, real-world exploration and realization of Einstein's theory of relativity. The desired accuracy comes, rather precisely, at the cost of fixed or absolute understandings of space and time.

Galison is of course talking about the Global Positioning System, the network of twenty-four military satellites that today helps everything from missiles to mobile phones know more or less exactly where they are on the face of the Earth: "The late twentieth-century GPS satellites provided precision timing (and therefore positioning) for both civilian and military users. Built into this orbiting machine were the software and hardware adjustments required by Einstein's theories of relativity. The result is a planet-encompassing, $10 billion theory machine."[11]

GPS, Galison says, unhinges our sense of stable and fixed location: "so accurate had the system become that even 'fixed' parts of the earth's landmass revealed themselves to be in motion, an unending shuffle of continents drifting over the surface of the planet on backs of tectonic plates." This "relativization" is not only a result of the unprecedented accuracy of the new measuring technology, however. It is also embedded in the very way in which it works. The system functions only because it takes this relativity into account in its timekeeping: "According to relativity, satellites that were orbiting the earth at 12,500 miles per hour ran their clocks slow (relative to the earth) by 7 millionths of a second per day," and "eleven thousand miles in space, where the satellites orbited, general relativity predicted that the weaker gravitational field would leave the satellite clocks running fast (relative to the earth's surface) by 45 millionths of a second per day." When corrections for these relativistic errors were built into the system, it worked: "relativity — or rather relativities (special and general) — had joined an apparatus laying an invisible grid over the planet. Theory had become a machine."[12]

But what kind of theory? Galison limits his claims to Einstein's theory of relativity, but he draws radical conclusions nonetheless. Einstein's theory, he argues, "designed a machine that upended the very category of metaphysical centrality. Absolute time was dead. With time coordination now defined only by the exchange of electromagnetic signals, Einstein could finish his description of the electromagnetic theory of moving bodies without spatial or temporal reference to any specially picked-out rest frame, whether in the ether or on earth. No center remained."[13]

In fact, GPS and a whole new set of technologies linked to it have introduced, or hyperbolized, a profound decentering or disorientation, and it is that loss of absolute reference points—and the political engagements and commitments that can be *enabled* by that loss—that are explored in the projects chronicled here.

FROM THEORY TO PRACTICE

We constantly read maps. In print and on computers, mobile phones, PowerPoint presentations, and blogs, maps visualize everything from the movement of hurricanes and refugees to the patterns of traffic and shifting electoral landscapes. Maps and the sophisticated technologies that create them are not limited, of course, to the public domain—we can only imagine the complex maps housed in the nose cone of a cruise missile or those that detail the location of every phone call and email intercepted by the Department of Homeland Security. But we tend generally to reduce maps to the diagrams we hold in our hands. They show us where we are and how to get somewhere else, and in doing so, they can contribute to a sense of security and self-possession. The solidity and certainty of the phrase "You are here" would be the motto of that identity-reinforcing—and maybe even identity-constitutive—function of maps.

The more they become our everyday means of navigating simple and complex situations alike, the more we take maps for granted. Rather than the interpretations of information that they are, we too often see them simply as representations and descriptions of space. This makes the task of analyzing them even more critical.

Maps locate. We can read them because they come laden with conventions, ranging from their legend, scale, and codes of graphic representation to what counts as the information they represent. They depend on a system of notation or of coordinates that places things in relation to one another.

This holds for maps that claim to represent physical spaces as well as those that diagram or chart the relative location of nonphysical entities: maps of a family or kinship structure, for instance, or the flows of data through a network. The spaces that maps try to describe can be ideal, psychological, virtual, immaterial, or imaginary—and they are never *just* physical.

This drive to locate, to coordinate, however revelatory and even emancipatory it can be, also has its price. It seems as though in the end, maps—the successful ones, the ones that show us where we are and get us from here to there—risk offering only two alternatives. They let us see too much, and hence blind us to what we cannot see, imposing a quiet tyranny of orientation that erases the possibility of disoriented discovery, or they lose sight of all the other things that we ought to see. They omit, according to their conventions, those invisible lines of people, places, and networks that create the most common spaces we live in today.

It is this comfortable sense of orientation, of there being a fixed point, a center from which we can determine with certainty where we are, who we are, or where we are going, that the projects in this book challenge. They put the project of orientation—visibility, location, use, action, and exploration—into question, and they do so without dispensing with maps.

The maps here are built with GPS, satellite images, databases, and geographic information systems (GIS) software: digital spatial technologies originally designed for military and governmental purposes such as reconnaissance, monitoring, ballistics, the census, and national security. Rather than shying away from the politics and complexities of their intended uses, these maps attempt to understand them. Poised at the intersection of art, architecture, activism, and geography, they intend to uncover the implicit biases of the new views, the means of recording information that they present, and the new spaces they have opened up. These projects expose the materials they work with in order to reclaim, repurpose, and discover their inadvertent, sometimes critical, often propositional, uses. They can be used to document, memorialize, preserve, interpret, and politicize, or simply as aesthetic devices, but as with all maps, the ones here—as well as the data sets and the technologies used to chart them—are not neutral.

"WHAT IS CALLED REALITY IS CONSTITUTED IN A COMPLEX OF REPRESENTATIONS"

Every spot on earth can be located, calculated, and represented in multiple descriptive systems. The digitization of the globe was prefigured by the ancient Greek system of latitudinal and longitudinal lines, translating the surface of the Earth into an abstract and universal grid. Irrespective of politics, place names, borders, or changing environments, places were fixed within the mathematical descriptions of their location.

A network of atomic clocks, cameras, and computers has built a virtual globe on which any point of physical space is easily coordinated with digital space. With this change comes the potential to move digital information very quickly from one place to another. We are familiar with the idea that new spaces are today being constructed—spaces different from the ones in which our bodies normally move—but we don't quite know what to think about them. They are the netherland spaces of electronic money, information warfare, and dataveillance, but they are also the spaces of the everyday, such as mobile phone calls, radio stations, navigation systems, and online social networks.

To call this the "coordination" of physical space with digital space, as I just did, perhaps understates things. The digital and the physical globes interact in profound ways, constituting in effect a question about which globe has the priority. In these days when virtual coordinates direct missiles to their targets and social networks have allowed phone companies and other collectors of our data trails to predict our next move in physical space, the shift has resulted in a radical transformation—we can never be sure which coordinate system takes priority in terms of representing our identity or our spatial movements.

Some years ago, Rosalyn Deutsche noted that "what is called reality—social meaning, relations, values, identities—is constituted in a complex of representations." This book experiments with that claim, tests its bearing on our new digital spatial realm, and ends up confirming it in its most radical formulations:

Reality and representation mutually imply each other. This does not mean, as it is frequently held, that no reality exists or that it is unknowable, but only that no founding presence, no objective source, or privileged ground of meaning, ensures a truth lurking behind representations and independent of subjects. Nor is the stress on representation a desertion of the field of politics; rather, it expands and recasts our conception of the political to include the forms of discourse. We might even say that it is thanks to the deconstruction of a privileged ground and the recognized impossibility of exterior standpoints that politics becomes a necessity. For in the absence of given or nonrelational meanings, any claim to know directly a truth outside representation emerges as an authoritarian form of representation employed in battles to name reality. There can never be an unproblematic—simply given— "representation of politics," but there is always a politics of representation.[14]

Representation and the Necessity of Interpretation

In 1977, the Eames Office, founded by Charles and Ray Eames, made a film called *Powers of Ten*. They aimed to explain "the relative size of things in the universe" by way of a sequence of images, zooming out in a series of frames from the aerial view of an unremarkable event, a couple having a picnic on a lawn, to the Milky Way and then back to a microscopic view of DNA.[15]

Citing the architect Eliel Saarinen, the Eameses argued for "the importance of always looking for the next larger thing—and the next smaller." This profoundly relativistic view animates their film about scale and the aesthetics of sliding along a scale; it is subtitled *About the Relative Size of Things in the Universe*. *Powers of Ten* constructs a seamless zoom into outer space, moving farther and farther away from the ground until the Earth becomes a tiny point in a much larger universe. Beginning with what we might call the human scale—the man and woman lying on a picnic blanket—the sequence of images reduces them (and their scale) to invisible insignificance, then reverses direction, returns to the surface of the Earth and its inhabitants, and then proceeds farther, all the way to the symbolic double helix of a DNA strand. "With a constant time unit for each power of ten," Ray Eames writes, "an unchanging center point, and a steady photographic move, we could show 'the effect of adding another zero' to any number." This steady move was what filmmakers Philip and Phyllis Morrison called "a disciplined smooth flow," "a long and uninterrupted straight line."[16]

The film intends to demonstrate that the universe is constructed as a set of transparent pictures, homogenous and continuous, telling more and more about its relational scale. In fact, however, the film tells us about the techniques of taking pictures of the Earth, its features and its context, at different scales. The zoom is *simulated* in the Eames movie, using more than a hundred separate images, many obtained from scientists and from NASA, others made in the studio, some even drawn and painted by hand.[17]

In a way, the apparently uninterrupted flow of the film, its seamless transition from one scale to another, might be seen as an attempt to compensate for its radically disorienting premise: There is no absolute scale, just as there is no natural or logical starting or stopping point for the zoom. It is not anchored anywhere—least of all in the human scale. Every scale is relativized by its proximity to and distance from the next, and there is no base or ground for the process of zooming itself. In the zoom we can see reaffirmed, even literalized, what Galison called the "upending" of "the very category of metaphysical centrality." The Eames's use of powers of ten as "an unchanging center point" was actually an exercise in radical decentering.

It took the Eames Office a long time and a lot of work to construct their zooms. Today, a nearly real-time zoom from the whole Earth to a picnic blanket is available on our desktops. And with a very easy interface, almost anyone can look at almost anything—not just a sentimental summer scene. The upending of the category of metaphysical centrality now is an everyday experience.

Today, "Google Earth" barely even names an application and its associated database; it is more of a nickname for our access to images of anyplace on the globe. Although it appears as a smooth zoom, the overhead view in Google Earth is just as much a composite, in its own way, as the "steady photographic move" of *Powers of Ten*. Instead of a comprehensive blanket of uniform-resolution (or real-time) images, it is a patchwork of archived aerial and satellite images of varying origins, sources, motivations, and resolutions. In fact, Google generates no overhead images of its own, but rather accesses them indirectly through the commercial enterprises that operate imaging satellites and via the people and governments who have tasked the satellites to collect data about specific locations at particular times. Google then assembles a composite map of these images, regardless of origin or resolution. For some places on the globe, Google Earth even has its own "archeological" record of the history of images of the spot, if and when those are available in the satellite company's database, and so it becomes possible to move backward and forward in time, as well as almost everywhere on Earth in space. Since 2008, by virtue of pressure from satellite image providers, Google also includes the name of the satellite company that has taken the picture.

How has this come about? The ease with which we can conduct these experiments often hides the reasons for the existence of the images in the first place. Why are they in the database, anyway? How did they get to be freely viewable online from 2005 on? The consumers of generally available satellite imagery, or even the ones who download images for a price from a commercial satellite database, will never know who has tasked a satellite to take a picture (unless they did it themselves) in order to see something close up, but from far away. And every view from a satellite is an experiment with the technology of looking close up at a distance, remotely examining and representing something as small as fifty centimeters of the ground from a height of four hundred miles in the sky.

In the ease of the Google Earth interface, as in the simplifications of a map, the political, military, and economic stakes that underwrite the creation and expansion of the database can often disappear. All that's left are the minimal data: the image has a date, a time stamp, and a series of coordinates in which it has been registered and made available for purchase by others, including Google Earth.[18]

Thus, when we use the ubiquitous zoom of Google Earth to look at our houses or neighborhoods, how many of us stop to consider that the image of our backyard was almost impossible to see—either because the image did not exist or its technology of the zoom was a military secret—only a short time ago? Moreover, how many people know what it is that they are looking at—a high-resolution commercial satellite image, a low-resolution one, or an aerial photograph?

The transition of satellite images from state secrets to commonplace everyday instruments that can be used for indulging idle curiosity, not just for implementing drone strikes on suspected terrorists, has been gradual, but accelerating. Only a few years separate the first Corona satellite mission (1964), tasked on high-resolution and top-secret image collection flights that were not declassified until thirty years later, and the launch of the first Landsat satellite (1972), a low-resolution environmental mission generating a potentially complete and publicly available world picture every three days. Some significant events mark the transition of satellite imagery from top secret to the taken-for-granted public availability that characterizes them now.

In August 1995, as a debate about mass killings the previous month at Srebrenica, in Bosnia, unfolded, Barbara Crosette reported in the *New York Times* that the Clinton administration had shown classified satellite and aerial photographs of mass graves and execution sites to the United Nations Security Council, but had made a distinction between them for the press.

> The administration made public three of the photographs, which showed disturbed soil, taken from a U-2 spy plane. It declined, however, to let reporters see the satellite photographs taken several days earlier, which were said to include pictures of people crowded into a soccer field. American officials said the satellite photographs were classified, although Secretary of State Madeleine Albright showed them to the other fourteen members of the Security Council.[19]

Thus, a residue of reticence and secrecy remained, for some images, just weeks after Vice President Al Gore had inaugurated one of the most ambitious declassification efforts in U.S. history with the unveiling of the CIA's Cold War–era Corona project and its extraordinary visual archive.[20] The example of the Srebrenica images, though, was the significant one: since then, we have lived in a geopolitical

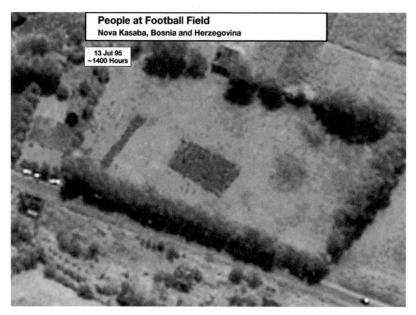

People at Football Field
Nova Kasaba, Bosnia and Herzegovina

13 Jul 95
~1400 Hours

U.S. satellite image taken on July 13, 1995, showing about six hundred people gathered in a soccer field at Nova Kasaba, Bosnia-Herzegovina, near Srebrenica. It was one of several classified images shown to members of the UN Security Council on August 10, 1995, as evidence of mass killings by the Bosnian Serb Army.

IMAGE: INTERNATIONAL CRIMINAL TRIBUNAL FOR THE FORMER YUGOSLAVIA, VIA U.S. HOLOCAUST MEMORIAL MUSEUM

world in which it was not only a reasonable working assumption that major events could be monitored from outer space, but that the traces of that surveillance would appear in the public sphere.

In 2000, the *New York Times* for the first time used the newly available Ikonos satellite as a sort of alternative investigative journalist in Chechnya. On the front page of the Sunday "Week in Review" section, two comparable satellite images of the Chechen capital city of Grozny were published, bearing the title "Campaign Poster." The first image was dated December 16, 1999, and the second March 16, 2000, just ten days prior to their publication in the newspaper. The accompanying text remarked on the likely electoral victory that day of Russian President Vladimir Putin and explained: "The images above, commissioned by the New York Times and taken by a commercial satellite, hint at the cost of that victory, in the destruction of a residential area near Minutka Square in the Chechen capital, Grozny."[21] As Lara Nettelfield has pointed out, "unlike other images of destruction in the post-Communist world, the Grozny pictures failed to arouse public sympathy or outrage for the plight of civilians in Chechnya."[22] But since then, this genre

Editorial and Op-Ed pages, 16-17
Education Advertising
Careers in Education and
Health Care Employment

𝕿𝖍𝖊 𝕹𝖊𝖜 𝖄𝖔𝖗𝖐 𝕿𝖎𝖒𝖊𝖘

Sunday, March 26, 2000

Week in Review

Section 4

Campaign Poster

Grozny, Dec. 16, 1999

Grozny, March 16, 2000

The *New York Times* first used before-and-after satellite imagery, directly obtained from a commercial provider, in an analysis of the role of the Chechen war in the Russian presidential campaign of March 2000.

of before-and-after images has become commonplace in much news gathering and reporting from zones of conflict and mass destruction.

Fast forward almost another decade. In April 2009, *The Lede* blog at the *New York Times* reported on what might be considered a satellite photo opportunity. "In what was either a remarkable coincidence or a bit of precision timing," wrote Robert Mackey, the North Korean government had launched a rocket "just as a commercial satellite, owned by a company which provides images of the earth to the Pentagon, DigitalGlobe, was passing over North Korea."[23] The *Guardian's* science correspondent, Ian Sample, reported that at least one British defense analyst

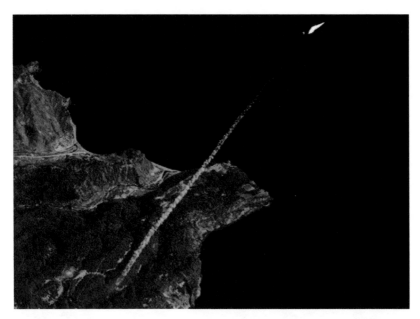

"A satellite image showing what is believed to be the exhaust trail and part of a North Korean rocket launched on April 5. The company that took the photograph, DigitalGlobe, describes it as 'a panchromatic, 50 centimeter (1.6 foot) high-resolution WorldView-1 satellite image showing the rocket launch from the Musudan Ri launch facility, formerly known as Taepo-dong.'" CAPTION: *NEW YORK TIMES*; IMAGE: DIGITALGLOBE

"suspects Pyongyang had timed the controversial launch to coincide with the satellite's arrival, in the hope of maximizing publicity of the launch."[24]

Google Earth is only the latest step in the public availability or democratization of high-resolution satellite imagery. Many military technologies have gone from classified to omnipresent, from expensive to free, and from centralized to distributed, downloadable on our desktops anywhere on Earth with access to the Internet.[25] That much seems certain. Policy analysts have dubbed this a "growing global transparency."[26] However, because what is involved is the appearance in the public sphere of a way of viewing things close up at a distance in which there is no absolute scale, no anchor, no center, evaluating this new visibility and negotiating its reality is a lot less obvious.

THE OPACITY OF TRANSPARENCY

In September 1999, Space Imaging successfully launched Ikonos, the first satellite to make high-resolution image data publicly available. John Pike, who pioneered the civilian use of aerial and satellite imagery at the Federation of American Scientists and today directs GlobalSecurity.org, called it "one of the most significant events in the history of the space age."[27] Earlier, Pike had suggested that a new kind of deterrence was enabled when news organizations and civilians could test, with meaningful certainty, the authority of official claims about, for example, the presence or absence of nuclear facilities in other states. And likewise, "it provides an independent check," he said, "on what the government is saying, for example about mass graves and other wartime atrocities in the Balkans."[28] Ann Florini, of the Carnegie Endowment for International Peace, argued that "on the plus side, governments and nongovernmental organizations may find it easier to respond quickly to sudden refugee movements, to document and publicize large-scale humanitarian atrocities, to monitor environmental degradation, or to manage international disputes before they escalate…. But, there is no way to guarantee benevolent uses."[29]

When U.S. Secretary of State Colin Powell made his infamous February 2003 presentation to the United Nations Security Council claiming to demonstrate that the government of Iraq was in possession of weapons of mass destruction, he presented a PowerPoint slide show that included a lot of satellite images, annotated to support his claims. "The facts speak for themselves," he said. "My colleagues, every statement I make today is backed up by sources, solid sources. These are not assertions. What we are giving you are facts and conclusions based on solid intelligence." And later he repeated, "Ladies and gentlemen, these are not assertions. These are facts corroborated by many sources, some of them sources of the intelligence services of other countries."

Later, he clarified his epistemology. He explained that the images, in fact, did *not* speak for themselves and were indeed hard to understand, but insisted that he was confident in his own ability, backed by the work of experts, to say what they meant:

> Let me say a word about satellite images before I show a couple. The photos that I am about to show you are sometimes hard for the average person to interpret, hard for me. The painstaking work of photo analysis takes experts with years and years of experience, poring for hours and hours over light tables. But as I show you these images, I will try to capture and explain what they mean, what they indicate, to our imagery specialists.[30]

The images he presented had been artfully interpreted, which is not to say that they were fake or forged or even that the images distorted the truth. Simply and more importantly, they were not objective photographs, but were presented as such. They were interpretations presented as facts and in a way that prevented anyone else from examining the uninterpreted data. The presentation and its cata-strophic results remind us that we need to be alert to what is being highlighted and pointed toward, to the ways in which satellite evidence is used in making assertions and arguments. We need to learn how to agree and disagree with those arguments, to challenge the interpretations made of images that are anything but objective or self-evident. For every image, we should be able to inquire about its technology, its location data, its ownership, its legibility, and its source. To facili-tate that inquiry, an image and its associated data should remain closely linked. But we are seldom given access to the data or the tools with which to interpret it, because the satellite images have been stripped of their data and presented to us as pictures already interpreted by experts.

We know now that there were no weapons of mass destruction found in Iraq. We also know that there was a videotape made by a jihadist militia, the Islamic Army in Iraq, that showed the group using satellite images from Google Earth to plan an attack.[31] And we have witnessed the "benevolent" stand taken by the Satellite Sentinel Project at Harvard, which makes use of DigitalGlobe satellite imagery to "identify chilling warning signs [of mass atrocities]—elevated roads for moving heavy armor, lengthened airstrips for landing attack aircraft, build-ups of troops, tanks, and artillery preparing for invasion—and sound the alarm."[32] Michael Van Rooyen, director of the Harvard Humanitarian Initiative, which houses the project, says that it's "a clear example of how technology transforms the way we think about and prepare for crises. In the hands of well-trained and experienced analysts guided by humanitarian principles, satellite technology pro-vides a potent new way of ensuring that the world witnesses threats to civilians."[33]

Is the globe transparent? Yes, sort of. High-resolution satellites seem to sig-nify global transparency, to realize effectively the dream that pretty much anyone could be able to see pretty much anything, anywhere. Because a visual regime that is inherently decentering, that disorients under the banner of orientation, can be used for all sorts of purposes, understanding how the images thus generated are produced and used is a civic responsibility and a political obligation. And the ways in which these satellite views are for the most part presented to the public—which is to say, in the news or in the public announcements of private companies, NGOs, or government agencies—are as misleading as they are revelatory: they come to us as already interpreted images, and in a way that obscures the data that has built them. As apparently self-evident images, pictures stripped of their data, they gen-erally lack, omit, or erase the fact, quite simply, that they have been interpreted.

In such a situation, Lisa Parks worries that any satellite image, even on Google Earth, implies a military view, which is to say, "knowledge practices of intelligence gathering and Earth observation...satellites...encircling the Earth on planetary

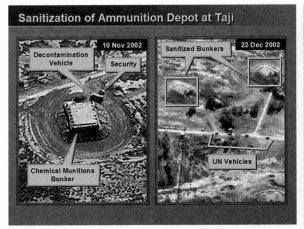

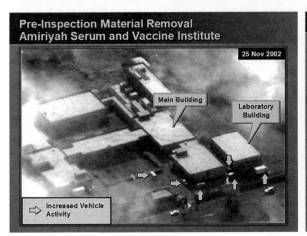

Four slides from U.S. Secretary of State Colin Powell's PowerPoint presentation on Iraq to the United Nations Security Council, February 5, 2003. These are some of the many annotated satellite images displayed that day, without access to the original satellite data.

IMAGES: U.S. DEPARTMENT OF STATE

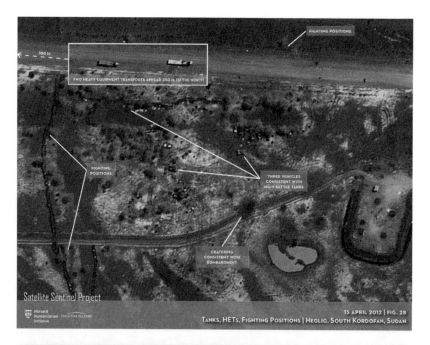

FIGHTING POSITIONS

350 M

TWO HEAVY EQUIPMENT TRANSPORTS APPEAR 350 M TO THE NORTH

FIGHTING POSITIONS

THREE VEHICLES CONSISTENT WITH MAIN BATTLE TANKS

CRATERING CONSISTENT WITH BOMBARDMENT

Satellite Sentinel Project

Harvard Humanitarian · DIGITALGLOBE Initiative

15 APRIL 2012 | FIG. 2B
TANKS, HETs, FIGHTING POSITIONS | HEGLIG, SOUTH KORDOFAN, SUDAN

PIPELINE

CRATERING CONSISTENT WITH BOMBARDMENT

CRATERING CONSISTENT WITH BOMBARDMENT

CRATERING CONSISTENT WITH BOMBARDMENT

Satellite Sentinel Project

Harvard Humanitarian DIGITALGLOBE Initiative

15 APRIL 2012 | FIG. 3
CRATERING CONSISTENT WITH BOMBARDMENT NEAR OIL INFRASTRUCTURE | HEGLIG, SOUTH KORDOFAN, SUDAN

"The Satellite Sentinel Project (SSP) has published new imagery indicating that as Sudan and South Sudan clashed over an oil field near the disputed border town of Heglig, a key part of the pipeline infrastructure was destroyed. The damage appears to be so severe, and in such a critical part of the oil infrastructure, that it would likely stop oil flow in the area, according to SSP."

FROM PRESS RELEASE, HTTP://WWW.SATSENTINEL.ORG; SATELLITE IMAGERY FROM DIGITALGLOBE

patrols" and "treat[ing] the surface of the Earth as a domain of unobstructed Western vision, knowledge, and control." She says, "I define remote sensing as a televisual practice that has been articulated with military and scientific use of satellites to monitor, historicize and visualize events on Earth."[34]

Parks spends a long time analyzing the same images of Srebrenica mentioned above, the ones Madeline Albright showed to the Security Council. Some of her analysis may seem apologetic now, for instance, her protests against the ways "the Western media tended to demonize the Serbs" and her suggestion that the Bosnian Army "must take partial responsibility for conditions leading to the massacre" at Srebrenica.[35] Her response is complicated by her almost automatic suspicion whenever agencies of the U.S. government appear to be the main interpreters of events by way of a satellite image. However, she is right that we were simply given images and interpretations by Albright, and more importantly, that we were also and unexpectedly seeing high-resolution intelligence imagery used for the first time as evidence of genocide. But today it is clear that Albright's imagery was in fact essential evidence of a crime and its cover-up.

Parks devotes a good part of her analysis to the "passive-aggressive voyeurism" of the U.S. government, "idly recording" the attack on Srebrenica while failing to do anything to stop it. She concludes that because of its "remoteness and abstraction," the satellite view functioned merely as an "overview of the war, draw[ing] on the discursive authority of meteorology, photography, cartography and state intelligence to produce its reality and truth effects."[36] The combination of passivity (just watching) and aggressivity (the militarized view) is most troubling to her. The problem is not just that the image comes from the state, though, and bears its codes; she seems troubled by its ontology, as well.

> Since it is digital, however, the satellite image is only an *approximation* of the event, not a mechanical reproduction of it or live immersion in it.... Because it is digital, its ontological status differs from that of the electronic image. The satellite image is encoded with time coordinates that index the moment of its acquisition, but since most satellite image data is simply archived in huge supercomputers, *its tense is one of latency*.... The satellite image is not really produced, then, until it is sorted, rendered, and put into circulation.[37]

. This latency or approximation, for Parks, leaves the satellite image open to all sorts of exploitation, most notably that operated by a military-diplomatic machine promoting its own omniscience and objectivity. Critical of that, she endorses the engagement of journalist David Rohde, who traveled to the scene in the immediate aftermath of Albright's revelations to see whether he could confirm what the images seemed to show. She admires his success in "witness[ing] the minutiae that the satellite could not pick up," his eyewitness account of the body parts, clothes, shell casings, and documents left behind in the mass murders. She is tempted by the notion that, because the eye sees at a higher resolution than the satellite, it sees more, and more clearly. Parks praises Rohde's "refusal to accept the satellite image as evident"—"instead of accepting the state's attempt to anchor the meaning of the satellite image, [he] seizes its emptiness and abstraction as impetus to infuse it with partiality, situated knowledge, and local tales."[38]

She leaves unstated the fact that Rohde in fact did confirm the interpretation that Albright had offered, but that is less important than her commitment to what is called "ground truthing." The ultimate test of the image, it seems, is what can be found, seen, heard, and sensed on the ground itself. In fact, she makes the trip herself some years later, but confesses to not really being able to see very much. Srebrenica was still largely populated by those who had killed and expelled its Muslim population and neighbors: "at Cafe Kum I encountered a former Serbian military officer who, I was told, had recently been indicted by the War Crimes Tribunal." So it was hard to learn much. "There is a code of silence in Srebrenica that is difficult to penetrate, especially for an outsider like me," she says. She concludes from this persistence of unreality—"the site was as abstract to me up close as when I first saw it on television"—that "witnessing became *a fantasy of proximity*."[39]

This conclusion seems more reliable than the premise that generated it. The view up close can be just as blurred as the one from overhead, and the difference between the image as a "site of activity" and a "memorial" more difficult to tell than it might at first seem.[40] What is most valuable here is the caution she invites: no satellite image presents a simple, unambiguous picture of the Earth, and a visit to the site itself can often raise more questions than it answers, reaffirming rather than reducing the openness of the image to interpretation. In the end, it seems, embedded in the very structure of the techno-scientific, militarized, "objective" image is something more disorienting, an "emptiness and abstraction" that resists sovereign control and opens itself to other sorts of interpretation.

INTERPRETATION AND "THE VIEW FROM NOWHERE"

What does the "emptiness and abstraction" of digital satellite images reveal? Although how such images are to be read and who is able to read them are of central importance, widespread understanding of satellite imagery and how to interpret it lags considerably behind its rate of production. We are often presented with images bereft of any data associated with them and subordinated to the interpretations that guard that data behind a shield of security and expertise. The projects in this book aim to challenge that.

What digital satellite images can show is certainly derived from a military or logistical worldview and deeply indebted to the institutions committed to seeing the world in military or logistical terms. The publicly available images in Google Earth come largely from DigitalGlobe and GeoEye, both major contractors for the U.S. government in the development and deployment of high-resolution satellites. By allowing transparency and openness—or rather, by funneling these images to the public via Google Earth—the United States has remained, thus far, in the forefront of viewing at high resolution across borders. And because what is at issue here is interpretation, other interpretations are possible.

What is largely missing from Parks's argument is the positive reading of what "image interpretation" implies. It is both an art and a science, especially with satellite image data, and the relation between the two is not an easy one to negotiate, even for "experts" whose expertise is at the service of governments and commercial institutions. John Pike, interviewed on National Public Radio about satellite imagery of destroyed villages in Darfur, responded to his host's claim that "the interpretation of these images is an art as well as a science" this way:

> Well, it's a discipline that the military intelligence community has spent a long time training people to do. One of the big challenges with this type of imagery is in finding things that it's readily understandable what you're looking at, and doesn't require any great leap of imagination, you're not dependent on somebody else captioning it. In the case of the Chinese nuclear submarine, well, that was pretty straightforward. In the case of Darfur, frankly, I've been very frustrated that the satellite imagery has not had the sort of impact on the public imagination that we had hoped it would in the past.[41]

Pike is telling us about the leaps of imagination that image interpreters must take when they look at an image, and longing for images that require fewer and shorter leaps. We continue, though, to defer to experts and to privilege the view that designates itself as scientific and objective.

But because the interpretation of such images is an art, as well as a science—because it inherently involves imaginative leaps—the putatively scientific and objective interpretations at the service of governments and commercial institutions tell only *a* story, not *the* story, of what is going on in these images. Views of the globe, which is to say, maps, have always combined the science of spatial description and documentation with a certain art, as well. J.B. Harley argued famously that maps should be understood as multidisciplinary artifacts, ones that reveal social and political forces, as well as representations of power. He worried, in 1989, about the ways in which "the scientific rhetoric of map makers [was] becoming more strident." "Many may find it surprising," he wrote, "that 'art' no longer exists in 'professional' cartography." He asks that we question the by now naturalized conventions through which maps have in fact standardized our images and knowledge of the world. He also asks us "to search for the social forces that have structured cartography and to locate the presence of power—and its effects—in all map knowledge." Although Harley's article was aimed at historians, *against* "what cartographers tell us maps are supposed to be," his questions are equally important for professional cartographers and the users of maps.[42]

He asks about the legends and frames of ancient maps, whose creators could only imagine what the globe looked like, as well as the symbols and legends in contemporary maps, which claim the status of objective description of reality. He treats both as texts that need to be read closely so we can start to understand the bias in any map projection. He reminds us that even something as simple and innocent as the mathematical translation of a sphere projected as a so-called undistorted flat plane has a "politics." "In our cartographic workshops we standardize our images of the world," he writes, and the process is complex: "the way maps are compiled and the categories of information selected; the way they are generalized, a set of rules for the abstraction of the landscape; the way the elements in the landscape are formed into hierarchies; and the way various rhetorical styles that also reproduce power are employed to represent the landscape." The standardized cartographic images to which we have grown so accustomed that most of us don't know them as a particular interpretive decision—the Mercator projection—are distinguished from others because they project the spherical globe as a series of apparently undistorted square shapes. This formal, but not only formal, gesture, he points out, "helped to confirm a new myth of Europe's ideological centrality."[43]

Svetlana Alpers attributes these standardized images of the world, or the flattening of the Earth into the mathematical uniformity of longitude and latitude, to a certain disappearance of the subject, or what, following Thomas Nagel, she calls "the view from nowhere."[44] As an art historian, she opposes this flat surface to the equally mathematical formula of the perspectival grid, which is viewed from somewhere—the point of view of the subject who both constructs and is constructed by that view. Perspective, it is well known, freezes a subject in a particular place and time.

Maps do not employ perspective. Although the grid that the Mercator and other such projections impose on the sphere of the Earth may share with perspectival paintings the mathematical uniformity of the frame and the definition of the picture as a window through which an external viewer looks, they do not share the positioning of the viewer. The cartographic projection is, in that sense, viewed from nowhere.[45]

Maps construct a spatial interpretation through their techniques of represen-tation, the "normalized" views that Harley decries.[46] A cartographic projection transforms, mathematically, a sphere into plane.

Yve-Alain Bois arrives at maps, although he does not quite specify that this is where his argument leads, from another type of constructed, measured, and projected view: the "axonometric" projection. An axonometric drawing shows an object in ways that cannot be seen simply by looking at it. To do so, it rotates the object along one or more of its axes such that the surfaces of the top and two sides are in view simultaneously. The horizontal and vertical dimensions are projected to scale, so that their planes are parallel to each other. Unlike in a perspectival draw-ing, there is no single fixed position from which the object is viewed.

Axonometric drawing originated, argues Bois, in techniques developed by engineers in 1822 to draw carefully the joints of a new material, iron. What distin-guishes this technique is that the top and the side views are both drawn to scale, as if one were flying over the joint, but no perspective is generated to distort the scale. The engineers, Bois writes in "Metamorphosis of Axonometry," derived their drawings from French military artists a century and a half earlier, who had used the technique to simulate the trajectory of a cannonball making its way over the walls of a medieval city, in order to compensate for the blindness imposed on them by the walls.[47]

Modern architects reinvented this drawing technique another hundred years later, in 1923, showing an object from the top and the side view in equal measures in order deliberately to generate a decentered modernist aesthetic of ambiguity. "All treatises which precede this event…regardless of their concern with architecture, military art, technical drawing or geometry, emphasize the convenience and accu-racy of axonometry, whereas modern artists celebrated its perceptive ambiguity…. The axonometric image is reversible; it tears free of the ground (Malevich's term), facilitating aerial views." After chronicling the various ways in which more and more architects, from Herbert Bayer to the New York Five, used the axonometric view to focus on ambiguous spaces, rather than to reproduce the scientific or fac-tual vision of the engineer, Bois pushes the argument further to propose that the "history of axonometry should include a chapter on aerial views and photogram-metry." And there is no reason to stop there: the history should extend to remote sensing in all its forms…a history precisely, as Bois insists, not only of the logisti-cally and pragmatically military, but also at the same time of instability, abstrac-tion, "ambiguities," and the "vertiginously ambivalent."[48]

"The axonometric drawing hovers or flies above its object," concludes Bois.[49] Denis Cosgrove has written some of the history of this flying image, focusing on Oskar Messter's 1915 invention of the airborne automatic camera, which "allowed pilots to film a 60-by-2.4 kilometer strip of land surface in a sequence of frames at the scale of conventional topographic maps." With it, he says, "a new mode of geographical representation was created: 'a flattened and cubist map of the earth,' which demanded new skills to relate the image to the ground": "Composite photographic images demanded a different way of looking than the still photograph did. The eye moves over the virtual space of the image as across a map, parodying in some measure the kinetic vision of the flyer.... Over time the aerial photograph and, more recently, remote-sensed images have become codependent with the map."[50]

Although high-resolution satellite images are by now naturalized as authoritative and maplike, the rigor (and we could even say the truth) of their embeddedness into the coordinates of longitude and latitude, the digital grid of navigational lines, should not be allowed to efface their military-political origins, or the technologies that have produced them, or the "relativity" and "ambivalence" that can render them so profoundly opaque and disorienting—and demanding of interpretation.[51]

PARA-EMPIRICISM

Not only is the physical surface of the Earth being mapped—*we* are also part of the transformation effected by digital mapping technologies. Anything that is listed, counted, and linked to a physical or digital address can potentially become spatial data and be mapped as well. Mounds of social, financial, and mobile data are collected on a daily basis by private and public entities, and we are being counted and translated into data each time we interact with electronic networks. Maps are being generated and updated constantly with this data. All of us—crossing a border, talking to a census taker, swiping a credit card, riding the London Underground, entering a luxury building in Dubai or a public housing project in Seattle, withdrawing cash at an ATM, driving through a highway toll booth—can become, and are regularly becoming, points on all sorts of maps. The social city is inscribed repeatedly onto the physical city.

The projects in this book use advanced digital technology and data. I have each time taken a leap and not left the data it to speak for itself, but have tried instead to offer a reflection on what can be done with it. When working with data, things are not as obvious as they might seem. So while others call working with data "quantitative," "empirical," or "objective" analysis, I prefer the somewhat more modest notion of "para-empiricism."

The English prefix "para" comes from the Greek word meaning "by the side of, beside," hence "alongside of, by, past, or beyond." It has come to denote, in words such as "paramedic" and "paramilitary," the sense of auxiliary, almost, not quite, functional but not really a substitute. It is with this double sense of alongsideness and incompleteness that I employ this neologism.

Usually when we appeal to *data*, we mean by this nothing less than reality itself, the concrete facts of the world, the real things. We ask for data points, we collect them in data sets and databases, and we treat them as indexical traces of the very phenomena we wish to understand or manipulate. Data are, in their etymological sense, the givens with which we can operate on the world. When empirical social scientists want to explore the hard facts of a situation, it is to data in this sense that they turn.

Instead, the word "data," in this book, means nothing more or less than representations, delegates or emissaries of reality, to be sure, but only that: not presentations of the things themselves, but representations, figures, mediations— subject, then, to all the conventions and aesthetics and rhetorics that we have come to expect of our images and narratives. All data, then, are not empirical, not irreducible facts about the world, but exist as not quite or almost, alongside the world: they are para-empirical.

To put it another way, there is no such thing as raw data. Data are always translated such that they might be presented. The images, lists, graphs, and maps that represent those data are all interpretations. And there is no such thing as neutral data. Data are always collected for a specific purpose, by a combination of people, technology, money, commerce, and government. The phrase "data visualization," in that sense, is a bit redundant: data are already a visualization.

My claim is not that this plunges us into some abyss of uncertainty, though, or makes it impossible to function in the real world. On the contrary, it is only on the condition of accepting this condition of data, in *para*-empirical condition, that we have any chance of operating responsibly in or on the world. It is because we admit that our data are not the same as reality, that there are disputes about data and that they can be decided only in debates with others, that the realms of politics and ethics open up for us.

Here I share the position of Bruno Latour, who argued in his introduction to the catalog of his ZKM show, *Making Things Public*, that the time has come for a thorough reevaluation of the so-called "crisis of representation." It might be, he says, and he means that this is in fact the case, that

> half of such a crisis is due to what has been sold to the general public under the name of a faithful, transparent and accurate representation. We are asking from representation something it cannot possibly give, namely representation without any re-presentation, without any provisional assertions, without any imperfect

proof, without any opaque layers of translations, transmissions, betrayals, without any complicated machinery of assembly, delegation, proof, argumentation, negotiation and conclusion.[52]

"Para-empiricism" names for me this effort at once to reclaim a sense of reality, and not to imagine that this requires doing away with representations, narratives, and images.

The projects included here don't only talk *about* maps, images, data. They seek to talk *with* them—to put them to use in ways that are critical of or that enlarge our conceptions of where we are and might be in the world. From the facts on the ground to the exhilaration of disorientation, the projects and writing, the images and data, collected here all aim to open spaces for discussion and action. They affirm the necessity of critique, and they reject the idea that critique requires "critical distance," at least in the ordinary sense. That is, they aim to make more space in the public sphere for the participation of everyone, not just governments, their militaries, and the experts tasked with making interpretations of global imagery to serve those constituencies. They aim to make it possible for everyone at least to understand how to participate actively, and by necessity politically, within the new territories constituted by these technologies of representation.

LEXICON

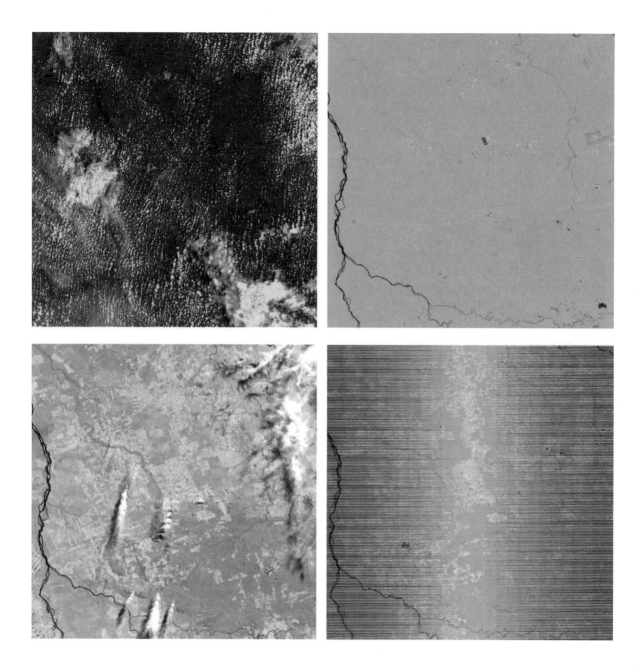

Landsat's forty-year archive gives researchers the ability to investigate changes at the same location over time. These images, acquired by Landsat 1, 4, 5, and 7, show an area near Nova Monte Verde, a municipality in the Brazilian state of Mato Grosso, just south of the Amazon, where rain forest is being replaced by ranching and monocropping. Top left: November 30, 1972; top right: August 5, 1986; bottom left: May 5, 2006; bottom right: August 4, 2012. The striping in the last image is due to failure of Landsat 7's scan line corrector.

From Military Surveillance to the Public Sphere

The discussions of the projects in this book refer to a number of technologies used in the process of mapping—GPS, remote-sensing satellites, and GIS. The projects make use of them in order to create new images or repurpose existing ones. But the history and politics of these technologies are at once obscure and important for understanding what's at stake in working with them. The following lexicon attempts to sketch the stories of the development of these technologies, their technical language, and their political and historical contexts. This chapter, which largely eschews explicit theoretical reflection, is designed both to document the increasing public access to these technologies and to lay the groundwork for the discussions of how they have been put to use in the chapters that follow. The list is not a complete one, but touches on most of the technologies with which I have engaged.

GLOBAL POSITIONING SYSTEM (GPS)

The GPS is a network of twenty-four satellites and five ground stations designed to provide to anyone carrying a portable receiver a highly specific determination of his or her location, anywhere, anytime, and in any weather.[1] The satellites, launched and operated by the U.S. military, are arranged in six circular orbits at an altitude of 11,000 miles, which makes it possible for at least four of them to be "seen" at one time by a receiver anywhere on Earth, and they constantly emit signals specifying their time and their own positions. A GPS receiver measures the time that the different signals take to reach it, and by comparing that with what it learns about where the satellite is, the receiver can calculate its own position. GPS location and time signals are freely available to anyone with a GPS receiver, including those embedded in other devices, such as mobile phones and cameras.

The research and launch period for the Global Positioning System began in 1973 and ended in 1991, when the program became operational just in time for the first Gulf War. The first experimental satellite was launched in 1978, the first satellite in the system was launched in 1989, and the full constellation of twenty-four satellites, also known as NAVSTAR by the Department of Defense, was completed in 1993.[2] GPS is now not only a household word, but a ubiquitous technology—what the official GPS website calls a "U.S.-owned utility"—used for everything from directing missiles to their target, to tracking elephants, to locating mobile phones and their users, to everyday navigating on land and sea, to hiking in the mountains, to recording the precise time of a financial transaction, to playing urban games using geotagging devices, and beyond.

Originally designed to provide accurate measurements of positions to within 100 meters, GPS is now capable of locating a position within 5 meters of accuracy. Not everyone, however, has always been permitted to make use of this degree of accuracy. When the system was launched by the U.S. military, it was designated a "dual-use technology," which meant that its features were also available for civilian use—but in an intentionally downgraded way. Originally it was governed by a policy known as "Selective Availability," which intentionally scrambled the highly accurate signals so as to reduce accuracy readings to 100 meters for civilian users. It was possible for civilians to improve the accuracy using a technique called "differential correction," which involved gathering additional readings from base stations at known locations within roughly three hundred miles (the area covered by one group of four satellites) and correcting the errors by measuring against the location of the base stations. This allowed, even in the early days of the system, position readings between 2 and 5 meters of accuracy.

Over time, the accuracy and availability of the GPS system has been affected less by the limitations or capacities of the technology than by a series of U.S. government policy decisions.[4] The first was the decision to activate the system in a two-tier manner, with different quality readings available to military and civilian users.

Only five years later, in 1996, President Clinton committed the United States to the continued maintenance and upgrade of the system and announced that it was his "intention to discontinue the use of GPS Selective Availability (SA) within a decade, in a manner that allows adequate time and resources for our military forces to prepare fully for operations without SA."[5] In May 2000, the SA program was abandoned, and fully accurate GPS readings are now publicly and freely available.

Today, according to the U.S. government's online GPS information page:

The GPS signal in space will provide a "worst case" pseudorange accuracy of 7.8 meters at a 95% confidence level. The actual accuracy users attain depends on factors outside the government's control, including atmospheric effects and receiver quality. Real-world data collected by the FAA show that some high-quality GPS SPS receivers currently provide better than 3 meter horizontal accuracy. [FAA data from early 2011 shows GPS SPS was often accurate to ~1 meter.] Higher accuracy is available today by using GPS in combination with augmentation systems. These enable real-time positioning to within a few centimeters, and post-mission measurements at the millimeter level.... The accuracy of the GPS signal in space is actually the same for both the civilian GPS service (SPS) and the military GPS service (PPS). However, SPS broadcasts on only one frequency, while PPS uses two. This means military users can perform *ionospheric correction*, a technique that reduces radio degradation caused by the Earth's atmosphere. With less degradation, PPS provides better accuracy than the basic SPS. Many users enhance the basic SPS with local or regional augmentations. Such systems boost civilian GPS accuracy beyond that of PPS.[3]

In 2004, President Bush created the National Executive Committee for Space-Based Positioning, Navigation, and Timing (PNT) and adopted a new national policy committed to modernization, sustainability, and maintenance of GPS as a free worldwide utility.

Over the past decade, the Global Positioning System has grown into a global utility whose multi-use services are integral to U.S. national security, economic growth, transportation safety, and homeland security, and are an essential element of the worldwide economic infrastructure. In the year 2000, the United States recognized the increasing importance of the Global Positioning System to civil and commercial users by discontinuing the deliberate degradation of accuracy for non-military signals, known as Selective Availability.[6]

The policy acknowledges the development of European-based PNT systems and supports standards of interoperability and compatibility so that they might rely on each other's infrastructure. The policy also endorses a more accurate version of the system for military use, but without SA. In 2010, President Obama reaffirmed these policies.

Other nations have begun putting their own PNT systems into place. In Russia, the system is called GLONASS and has been in operation since 1995. Galileo is a system being developed by the European Union and other partner countries and is planned to be operational by 2014. There are other regional systems being planned by China, India, and Japan.

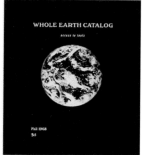

In 1966, Stewart Brand printed and sold buttons which asked the question, "Why haven't we seen a photograph of the whole Earth yet?" As his colleague Robert Horvitz wrote later, "Stewart wanted NASA to release a photo of the whole Earth because he believed it would have significant psychological impact: it would be visual proof of our unity and specialness, as our luminous blue ball-of-a-home contrasted dramatically with the dead black emptiness of space. Differences in skin color, religion, nationality and wealth, which can seem so important down here on Earth, shrink to nothing when viewed from afar." No spy satellite images were declassified. But a year later, NASA and a team of weather scientists at the universities of Wisconsin and Chicago released a film made of images taken by the newly launched ATS-III satellite in November 1967, titled "The First Color Movie of the Planet Earth: Viewed from 22,300 Miles over Brazil." And in the fall of 1968, the first issue of Brand's *Whole Earth Catalog* told readers how to buy a 16mm print of the film, and featured another image, also from the ATS-III spin-scan camera, taken over Brazil on November 10, 1967, on its cover. SATELLITE IMAGE: NASA

REMOTE-SENSING SATELLITES

The *Oxford English Dictionary* defines "remote sensing" as the sensing "of some-thing not immediately adjacent to the sensor; *spec.* the automatic acquisition of information about the surface of the earth or another planet from a distance, as carried out from satellites and high-flying aircraft." Remote sensing implies the collection of knowledge from an array of distances and methods, from human sight and sound to seeing and hearing from hundreds of miles in the sky or deep down in the ocean from the water's surface. What follows, however, focuses only on remote-sensing satellites and the technologies that allow us to see very closely from a distance.

Remote-sensing satellites have been launched since the 1960s, generally to an altitude of between 400 and 900 kilometers (249 and 559 miles) above the Earth, first by the United States and the Soviet Union (later Russia) and then by other states, including France, Israel, and India.[7] Remotely sensed images are generated either by the telescopic lenses of cameras or by sensors on the satellites. Older satellites captured what they sensed as analog images on physical storage surfaces, such as film, while later satellites have transmitted their data as digital information that is converted to images by ground stations. With either method, what remote-sensing satellites sense and record is reflected radiation: the ordinary visible light spectrum that allows us to see colors, and, since the 1970s, the nonvisible infra-red spectrum that allows, for example, for types of vegetation to be differentiated from each other by more than color.

This is all that each remote sensing satellite has in common. What follows out-lines a series of satellites used for remote sensing from 1960 until 2010. It is by no means a complete list, but can serve as an introduction to the satellites used here. The orbiting platforms range from spy satellites launched by the U.S. military and intelligence agencies (for instance, Corona), to those launched with public funds to monitor the Earth's resources (Landsat and SPOT), to privately launched satel-lites that today make very high-resolution imagery publicly available (for instance, Ikonos and GeoEye). This sequence tells the story of the technopolitical transfor-mation of access to remote-sensing imagery, a progression in both access and reso-lution that today makes very detailed images of the Earth from outer spac;e almost commonplace. The history is one of a tension between secrets and spying, on the one hand, and access and commerce, on the other, finally enabling nonprofession-als and civilians to make use of these powerful information resources.[8]

In my work, the satellites I have made use of are mostly those launched by the United States and operated by a combination of private corporations and U.S. government agencies. This is not an accident. Aside from the French SPOT satellites, launched in 1986, the United States has always had the highest-resolution imagery available and has maintained a set of policies designed to guarantee its global dominance in the field of satellite imagery.[9] This may change in the future. As with GPS satellites, other countries have launched high-resolution Earth-imaging satellites, including India, China, Japan, and Israel, and this list will expand to include Turkey, South Africa, and the Gulf Cooperation Council in the next decade.[10]

CORONA (UNITED STATES, 1959–1972)

Begun under the Eisenhower administration in reaction to the Soviet Union's Sputnik project, the Corona program focused primarily on photographing the Soviet Union and the People's Republic of China. The series of six classified satellites—dubbed KH-1 through KH-4B in a series of secret documents titled *Talent Keyhole*—produced high-resolution images for intelligence, reconnaissance, and mapping purposes. Today, Corona negatives and accompanying documents are available in the public sphere, prominently featuring the crossed-out words "TOP SECRET."

Over time, the ground resolution of Corona imagery improved from 40 feet to 5 feet.[11] Individual Corona images are film negatives, each recording 10 miles by 120 miles of ground space. The imagery was exposed on a newly designed physical polyester film, now known as Mylar. It was collected onboard the satellite in rolls and ejected or "de-orbited" in canisters inside a capsule with small parachutes, to be picked up in midair by aircraft at a location near Hawaii. "The capsules were designed to float, so that if the plane missed, Navy boats could retrieve them. In case the boats missed, the capsules were fitted with salt plugs that would dissolve after two days in the ocean, causing the capsule to sink beneath the waves, so the film could never fall into enemy hands."[12]

Rather than orbiting the earth for long periods of time, Corona satellites were "tasked" on missions to specific sites and territories. Corona was alternately used to spy on and to map certain locations. On its first successful one-day mission, August 18, 1960, KH-1 orbited the Earth only three times, taking pictures of 1.65 million square miles of the Soviet Union and Eastern Bloc countries on three thousand feet of film. Later missions lasted up to nineteen days, and the KH-4 satellites were equipped with two cameras—for both intelligence and mapping purposes. The last imagery was acquired by the KH-4B satellite on May 31, 1972. According to historian Keith Clarke, "The systems worked so well that in short order the CIA was using Corona to map the world, remap the U.S., and to evaluate all 1:24,000 topographic maps for revision."[13]

The archive of over eight hundred thousand Corona images—2.1 million feet of film in thirty-nine thousand canisters[14]—was declassified on February 22, 1995 with President Clinton's Executive Order 12951. The archive became available to the public three months later.[15]

LANDSAT (UNITED STATES, 1972–)

Appearing concurrently with the nascent environmental movement of the 1970s and dubbed the ERTS-1 (Earth Resources Technology Satellite), Landsat names a series of seven satellites launched by the National Aeronautics and Space Administration (NASA). The first was launched in July 1972. Together, they comprise the first publicly accessible remote-sensing program. Of these seven satellites, only Landsat 5 and Landsat 7 are currently functioning. A further satellite, known as the Landsat Data Continuity Mission, is scheduled for launch in 2013.[16]

Over time, ground resolution of the Landsat images has increased from 80 meters to 15 meters, which is officially described as "moderate." Each Landsat scene measures 170 by 185 kilometers (106 by 115 miles) of ground space. At its highest resolution, Landsat can picture large buildings and airstrips. According to a NASA presentation on Landsat, "this is an important spatial resolution because it is coarse enough for global coverage, yet detailed enough to characterize human-scale processes such as urban growth."[17] Landsat satellites orbit the Earth on predictable paths. The same coordinates are imaged at nearly the same time of day, every fourteen to eighteen days.

Because Landsat imagery is inexpensive and readily available, it is used frequently by researchers to investigate and highlight large-scale patterns related to climate change, natural resource management, land development, or disaster recovery. However, Landsat was not always so accessible. In the early 1980s, the program was privatized, and the National Oceanic and Atmospheric Administration (NOAA) selected the Earth Observation Satellite Company (EOSAT), later known as Space Imaging, to archive, collect, and distribute Landsat data as well as to build, launch, and operate the next two Landsat satellites (with government subsidies). As NASA tells the story today, "commercialization proved troublesome, with NOAA and EOSAT raising the cost of images by 600%, effectively "pric[ing] out many data users." Faced with competition from the newly launched French SPOT satellite and with coverage collapsing because EOSAT acquired imagery only when there were customers to buy it, Landsat images nearly disappeared by the end of the decade. "By 1989," reports the NASA Landsat history, the program was in such shambles that "NOAA directed EOSAT to turn off the satellites (no government agency was willing to commit augmentation funding for continued satellite operations, and data users were unwilling to make the hefty investments in computer processing hardware if future data collection was uncertain)."[18]

Over the course of the 1990s, control of Landsat's satellites and its imagery output was gradually returned to the U.S. government.[19] The pivotal role of Landsat imagery in the planning and implementation of the Gulf War, coupled with competition from the newer and cheaper SPOT, led to the Land Remote Sensing Policy Act, signed into law by President Clinton on October 28, 1992. It bolstered the Landsat program, stating that "continuous collection and utilization of land remote sensing data from space are of major benefit in studying and understanding human impacts on the global environment, in managing the Earth's resources, in carrying out national security functions, and in planning and conducting many other activities of scientific, economic and social importance."[20] The latest satellite, Landsat 7, was launched in 1999, and on July 1, 2001, operational control of the entire system and its archive was officially returned to the federal government, with EOSAT/Space Imaging giving up their commercial right to Landsat data. The program appears to be set to continue with the Landsat Data Continuity Program. Landsat images can be obtained from http://landsat.gsfc.nasa.gov.

SPOT (FRANCE, 1986–)

SPOT, an acronym for Système Probatoire d'Observation de la Terre, is a series of five satellites launched between 1986 and 2002 by the French national space agency, the Centre National d'Études Spatiales (CNES), in collaboration with Swedish and Belgian scientific agencies. At the time of its initial launch, SPOT 1 posed a serious challenge to the U.S. and Soviet monopoly on satellite imagery by offering 20-meter and 10-meter spatial resolution, significantly better than Landsat. Of the five satellites, SPOT 4 and SPOT 5 are currently functioning, and Astrium GEO Information Services (the private owners of the system) planned to launch two new satellites in 2012 and 2013 (SPOT 6 and 7) with ground resolution as high as 1.5 meters, as well as a successor pair of satellites called Pléiades, offering half-meter resolution (the first was launched in September 2012).[21]

Over time, SPOT image data has improved from 20 meters to 2.5 meters ground resolution at an altitude of 832 kilometers (517 miles). This resolution is able to capture small buildings, but not their details. SPOT orbits around the polar axis, capable of returning to the same place on Earth every twenty-six days.

In June 2010, the company announced a data-purchase agreement with the U.S. government allowing access to all image data collected by SPOT 4 and SPOT 5 over the United States. As with Landsat imagery (in partnership with NASA), the U.S. Geological Survey can distribute these images for free.[22] SPOT announced that its image data will therefore be the "most widely used medium resolution commercial sources of Earth observation data in the U.S. government."[23] This purchase may be the U.S. government's response to the pending danger in the Landsat data gap should a new Landsat satellite not be launched. Archival and recent can be purchased online through the SPOT catalogue at Astrium.[24]

IKONOS (UNITED STATES, 1999–)

Launched by the private company Space Imaging (the transformed EOSAT, now known as GeoEye) in September 1999, Ikonos-2 was the first satellite to make high-resolution satellite imagery available to civilian users, leading the *New York Times* to describe it some weeks later as "the world's first private spy camera."[25]

John Pike, then in charge of space policy at the Federation of American Scientists, told the *Times* that high-resolution imagery "was revolutionary when it was available to the nuclear powers, and one expects it to have similar potential now that it is commercial."[26] Robert Wright, writing in the *New York Times Magazine*, called it "a geopolitical milestone. Able to discern objects only a few feet wide—to see at 'one-meter resolution'—it will give presidents, generals and assorted political actors around the globe a kind of power once confined to elite nations."[27]

Ikonos was launched with the capability of providing image data with 1-meter ground resolution in a swath 11.3 kilometers (7 miles) wide from an altitude of 681 kilometers (423 miles). It functions by combining 82-centimeter (32.28-inch) resolution black-and-white ("panchromatic") images with 4-meter (13.12-foot) resolution multispectral images to create 1-meter (3.28-foot) color imagery (pan-sharpened).[28] At 1-meter resolution, Ikonos can distinguish a tank from a truck. Every point on Earth can be revisited by Ikonos every three to five days. Although its lifespan was a proposed seven years, Ikonos is still functioning.

Ikonos does not collect a steady stream of images. Its sensors are turned on only to record image data when tasked. Once the satellite is assigned an objective and the image data is received by a purchaser, it becomes available for repurchase and can be ordered and received through a website that includes the image data's longitude, latitude, and date stamp, but not the identity of the tasking agency or individual. Between its launch in 1999 and mid-2011, Ikonos had imaged more than 280 million square kilometers (over 100 million square miles) of the Earth's surface.[29]

The simultaneous provision of high-resolution image data to civilians, the U.S. military, and other governments globally was made possible by President Clinton's March 10, 1994, Presidential Decision Directive, which "among other things, loosened restrictions on the sale of high resolution imagery to foreign entities."[30]

According to the European Space Agency, "the spacecraft operations of Ikonos-2 are unique among the current commercial imaging satellites in that they allow each international affiliate to operate its own ground station(s). These ground stations are assigned blocks of time on the satellite during which they can directly task

On the front page of the *Washington Post* on March 3, 2005, Dana Priest revealed the existence of a secret CIA prison, code-named the Salt Pit, near Kabul, Afghanistan. Eight months later, she reported that the Salt Pit had been an early part of a "hidden global internment network," a series of so-called "black sites," in which the CIA housed and interrogated terror suspects. Her first article had offered enough detail to send GlobalSecurity.org looking for earlier satellite images of the Salt Pit, and so the second article included a high-resolution Ikonos satellite image of the building.

Top: Salt Pit, as seen by Ikonos satellite, January 25, 2001.   COURTESY GEOEYE

Bottom: Salt Pit, as seen by Ikonos satellite, July 17, 2003.   COURTESY SPACE IMAGING MIDDLE EAST

Ikonos, and immediately receive the downlinked imagery for which they tasked."[31]

The launch of Ikonos allowed the United States to retain its position as the primary provider of highest-resolution image data globally, but in so doing, it introduced sensitive issues of "shutter control," which, in the words of former Space Imaging vice president Mark Brender, "provides a lever by which the U.S. government can interrupt service when there is a 'threat to national security or foreign policy concern.'"[32] Rather than exercising shutter control, however, the U.S. government has deployed other means of controlling imagery during sensitive times: for example, purchasing the rights to all Ikonos image data over Afghanistan and Pakistan for the two months directly following the September 11, 2001 attacks on the United States. Images from the Ikonos archive, as well as new (tasked) acquisitions, are available for purchase worldwide through GeoEye.

QUICKBIRD-2 (UNITED STATES, 2001–)

QuickBird-2 was launched in October 2001, less than a year after the loss at launch of its predecessor, QuickBird-1. It is a high-resolution Earth-observation satellite owned by DigitalGlobe. It operates in a polar orbit, 482 kilometers (299.5 miles) above the Earth, with a swath width of 18 kilometers (11 miles). It is capable of sub-1-meter resolution, as high as 65 centimeters (25.6 inches).[33] Like Ikonos, QuickBird does not collect image data unless tasked to do so. It can revisit some sites beneath its orbit as frequently as every two and a half days, others within no more than six days. QuickBird-2 is also subject to shutter control, although the U.S. government has never implemented it.

QuickBird-2 and the other satellites in what DigitalGlobe calls its "constellation of sub-meter spacecraft" have emerged as major providers of overhead image data to the U.S. government. In a 2002 memo to the director of the National Imagery and Mapping Agency (NIMA), then–CIA Director George Tenet specified that "it is the policy of the Intelligence Community to use U.S. commercial space imagery to the greatest extent possible" and that the U.S. government should use commercial satellites unless military ones provide better resolution with classified image data.[34] DigitalGlobe has since been awarded two contracts by the U.S. government: $500 million from the NextView program in September 2004 and $3.5 billion over ten years from an EnhancedView contract in August 2010.[35]

The "sub-meter constellation" also does nongovernmental work. DigitalGlobe has agreements with humanitarian and human rights initiatives, among them the Satellite Sentinel Project at Harvard University, to provide QuickBird-2 and other images of zones of conflict in nearly real time. In March 2011, for instance, a DigitalGlobe vice president announced, on the company's blog, the release of satellite images of burned and destroyed villages in the Abyei region of Sudan. He wrote: "We've collected, processed, analyzed and delivered imagery and information in record time, given the urgency of the situation and the need to demonstrate to both sides that the world is watching." He added, for context, that this was simply part of the satellite business:

> we do keep a constant eye on the planet, to gain early insights into the business, market, environmental and political changes that impact people around the world. That's why we are keeping such a close eye on Sudan. It may be hard to watch, to look at an image and know someone's home is gone, a livelihood destroyed, that many lives have been lost. All involved are seeking the truth in pictures, and delivering valuable information and insight to both sides of the country. We certainly hope that one day, peace will come to this nation.[36]

QuickBird-2 image data can be purchased at digitalglobe.com, along with that of its fellow DigitalGlobe satellites WorldView-1 (50-centimeter/19.7-inch resolution) and WorldView-2 (46-centimeter/18.1-inch resolution).

GEOEYE-1 (UNITED STATES, 2008–)

The revolution in the privatization of high-resolution imagery from outer space that is exemplified by the generation of satellites from Ikonos on stems both from the declassification efforts of the 1990s and a series of U.S. government decisions to "support the continued development of the commercial satellite imagery industry by sharing the costs for the engineering, construction and launch of the next generation of commercial imagery satellites."[37] One result was the September 2008 launch of GeoEye-1, a private satellite owned by GeoEye with resolution below a half meter (41 centimeters, 16.41 inches). Its swath width is just over 15 kilometers (9 miles), and from its sun-synchronous polar orbit 681 kilometers high (423 miles), it can revisit anywhere on Earth once every three days, passing overhead, like other imaging satellites, at 10:30 a.m., local time. Like Ikonos, also owned by GeoEye, and QuickBird-2, it is subject to shutter control and does not collect imagery unless tasked to do so.

According to GeoEye, "While the satellite collects imagery at 0.41-meters, GeoEye's operating license from the U.S. Government requires re-sampling the imagery to 0.5-meter for all customers not explicitly granted a waiver by the U.S. Government."[38] Nevertheless, at this reduced 50-centimeter resolution, the home plate of a baseball diamond is visible from space.

GeoEye's CEO wrote in January 2010:

> The defense and intelligence communities have developed a huge appetite for unclassified, high-resolution, map-accurate satellite imagery. One leading reason is that our government can freely share unclassified images with allies, coalition

partners and disaster relief workers, thus speeding collaboration and time-critical decision-making. Another reason is that commercial imagery is highly cost-effective because we can resell excess capacity and imagery to commercial customers.

As a result, the use of satellite imagery by analysts and mapmakers at military headquarters is the norm.[39]

After the U.S. government, GeoEye's second major customer is Google. Since mid-2009, a lot of GeoEye-1 imagery has been freely available to Google Earth users. Although the Google logo was prominently displayed on the launch rocket—such that *Wired* magazine could title an article "Google's Super Satellite Captures First Image"[40]—Google does not own the satellite. Instead, through its Google Earth interface, it distributes and makes accessible imagery produced and tasked by others. (There is of course a possibility that Google has commissioned GeoEye imagery collection for its own purposes, but if so, it's a closely held secret.) It is unclear whether Google displays the GeoEye imagery at its full resolution, and since one cannot download images from Google Earth in the same way as one can from GeoEye itself—where each pixel has a size of one square meter and a longitude, latitude, and spectral signature—it's rather difficult to find out. For its full resolution and data, GeoEye-1 image data can be purchased at www.GeoEye.com.

GEOGRAPHIC INFORMATION SYSTEMS (GIS)

The Global Positioning System and remote-sensing satellites simply generate data. GIS is the generic name for software that allows users to locate data spatially. Any line on a spreadsheet, item on a list, or field in a database that records a physical address has the potential, once linked to its geographic coordinates, to become a point on a digital map. Once that point is recorded, it can be linked to or labeled with any other sort of data: the address can be connected to the name of a road, a dollar amount, a color or a shade, something a person said, a crime committed or thwarted, an encounter with an animal or a deity, or almost anything else that can be stored in a database—and that includes nonquantitative data.

Environmental Systems Research Institute (ESRI) is the Microsoft Word of GIS software and has the generic Web domain name www.GIS.com. GIS is described by ESRI as a system that "integrates hardware, software, and data for capturing, managing, analyzing, and displaying all forms of geographically referenced information."[41]

The most popular textbook on GIS, *Geographic Information Systems and Science*, describes the "field of GIS as concerned with the description, explanation, and prediction of patterns and processes at geographic scales. GIS is a science, a technology, a discipline and an applied problem solving methodology."[42] The textbook description says nothing about hardware and software, and rightly so, because it focuses on how GIS has radicalized and transformed the methodologies and processes of cartography, geography, urban planning, urban design, data management, archeology, sociology, and public health, among many other fields and practices. Although these are very different disciplines, they all have a stake in using maps as a basis for research and analysis.

Over the course of its short history, GIS has been commonly talked about as having transformed cartography into spatial data management. GIS has become a metaphor for the role that data now play in the drawing of new maps of the world, especially its cities and its resources. Often, the data is newly created for the map. What GIS does well is to layer diverse sets of information onto a single digital file or map.

Both these things—data displayed on maps and a layering of data onto maps—have long histories. Depending on where one starts the historical trajectory, one will end up with a very different interpretation of the meaning and uses of GIS. For example, some urbanists and public health researchers put the origin of GIS in John Snow's 1854 map of cholera in London. For them, the social data and statistical methods embedded in GIS are critical to the ways in which they define it.[43] These methods, which were developed later by Charles Booth in his poverty maps of London in 1898–99 and then by the Chicago School of sociological research in the first half of the twentieth century, constitute in effect the history of the modern city and define the modern history of cartography.[44]

But there are other genealogies. Some cite Ian McHarg's 1969 *Design with Nature* as the origin of GIS.[45] McHarg famously produced manually layered topographical maps with multiple sources of information in order to suggest ecologically smarter layouts for highways.[46] Slope, surface drainage, scenic value, residential values, forests, institutions, erosion, and so on were layered together into what McHarg called a "composite," an image in which one could see the effects of the layers on one another. The overlays bore titles such as "Composite: All Social Values" or "Composite: Physiographic Obstructions." McHarg's maps featured proposals such as "Recommended Minimum-Social-Cost Alignment" for a highway construction project. McHargian users of GIS have an expanded and design-oriented view of the built environment, one that incorporates ecological, landscape, and urban patterns, as well as the social forces that might affect those patterns.

The dominant history of GIS traces only the hardware and software that make up the GIS we know today on our computer desktops. The history section of *Geographic Information Systems and Science* begins in the mid-1960s in Canada, where the first "real GIS" was a "computerized map measuring system."[47] It was produced to create the Canada Land Inventory System, a project—classically cartographic—to identify resources and their potential uses.[48] A second phase of rapid development, they write, came from the U.S. Census Bureau, which, planning for the 1970 census, created the DIME (Dual Independent Map Encoding) program, allowing the creation of digital records of every street in the United States such that the population could be referred and aggregated to specific geographies. From the perspective of emerging GIS software development, these two programs responded to the "same basic needs in many different application areas, from resource management to the census."[49]

These narratives and genealogies are important as examples (and this is not the full scope of genealogical narratives of GIS) because neither data collection nor software are neutral in the uses of GIS. Sociologists, urban planners, advocacy groups, and other users of GIS software often tend to downplay the art of mapping and can unknowingly, or knowingly, as Mark Monmonier has argued, "lie with maps."[50] GIS software, which hides from the viewer or user of the map the statistical operations that the maker of the map utilizes, can make this traditional possibility a great deal easier. A more polite term for this, which acknowledges the explicitly aesthetic operations of some GIS users and recognizes the deployment of maps for persuasive purposes, as well as for the management of people and things, would be that of Dennis Wood, "the power of maps."[51]

Obviously, the design of the data and the reasons for its collection have an effect on the biases of the map. Now that many specialists other than cartographers can make maps, it is especially important to understand the sources of data they rely on, the products of which are maps and images that are having an effect on policy, cities, landscapes, privacy, and beyond.

Remote sensing had an enormous influence on the data and imagery in GIS. Aerial exploration of the Earth's surface not only generated the image bases for all sorts of maps, but also allowed interpreters to discover new things about everything from land use to population density to changes in landscapes and landforms. The Corona program was already using satellite imagery to map large parts of the United States and elsewhere by coordinating its measurable images with mapping reference grids (longitude and latitude). And as the 1990s dawned, GPS emerged as an unprecedented and inexhaustible source of new data points.

However, no one, really, would be using GIS were it not for the emergence of desktop and then portable computers and the World Wide Web, which dramatically democratized the availability of data-processing power in the late 1980s and early 1990s and effectively put GIS–like data and software into mass circulation.

With the ubiquity of personal computers and the increased availability of GIS software and geospatial data—whether from GPS, remote-sensing satellites, or public and private libraries and archives—the ability to access, interpret, and put to use digital images of events occurring anywhere in the world, on any scale, from the local to the global, is no longer the sole property of governments, militaries, and large corporations. What the dissemination of these technologies has enabled is the democratization of what I have called "para-empirical" investigations. What follows here are nine such investigations, together with reflections on the ways in which they can help us understand better how the images generated by this hardware and software are used, how the rest of us can explore their unintended consequences and unexpected byproducts—and how sometimes we can make such images ourselves.

PROJECTS

Each project in this book represents a snapshot in space and time taken somewhere between 1990 and 2010. Each revolves around an image or a set of images that emerged from the new geospatial technologies of those decades. The analytic dimensions of the projects have grown out of the work done on or with those images. The projects engage with newly accessible civilian spatial imaging technology, whether in the form of recently declassified military surveillance images, archived commercial satellite images, specially commissioned satellite imagery from newly launched satellites that were tasked as part of the project, GPS location data, or politically charged demographic data sets. In each case, the material seemed to me urgently in need of repurposing, recontextualization, interrogation, and re-presentation for ethical, activist, and memorial reasons.

Each of the nine map sections is presented here alongside two or three sets of texts. The first section consists of an after-the-fact commentary and reflection on these works, written retrospectively in 2012. The second section consists of the text (or portions of it) that accompanied the maps' original display or publication. These texts are archival records—meant to highlight the context in which the work was first presented and received—and barometers of the culture and the technologies with which they conversed. When there is a third section, it adds material that extends the original project.

The new façade of the Storefront for Art
and Architecture in Manhattan, designed
by Steven Holl and Vito Acconci, 1993.

PHOTOS: PAUL WARSCHOL

1 You Are Here

Actually to inhabit an information system

New York, 2012 — The Global Positioning System, the U.S. military's network of location satellites, was developed just in time to become fully operational at the onset of the first Gulf War as the American armed forces and a global coalition of allies went to war in and around and over Kuwait and Iraq. GPS–guided weapons flew through the air of the battlefield and across our television screens. And the GPS signals were everywhere—it was just a matter of finding a receiver, not so easy then, that could tune them in. The military and civilian versions of the technology appeared simultaneously, captivating experts and the public with the promise of the next utility—we would never be lost again. And yet very few people understood GPS. The *You Are Here* installations charted some of the territory opened up by—and in—this new and unfamiliar mapping system.

When I finally obtained a GPS receiver and its separate mushroom-shaped antenna, a loan from the manufacturer, Trimble, they came together in a small suitcase. The receiver itself was the size of a shoebox. Separate software, donated by Geolink, allowed the data to be displayed on a computer monitor in real time.

I was interested in the mutability of digitized information. Data sets are subject to a series of presentational choices, and points in space can be described in different ways: numeric, linguistic, or symbolic. They can appear as singular points or be connected into lines. In *You Are Here*, I chose to keep the points as points and to represent them as crosshairs, rather than as dots. The AutoCad crosshair became the navigational insignia of this research.

This project afforded the opportunity actually to inhabit an information system: to find that the totalizing system of the network itself is scattered, even in its principal architectures—to discover at the heart of the ideology of precision not a space of certainty, but one of confusion.

The first installation I did with GPS was *You Are Here: Information Drift*, the second show behind the Steven Holl/Vito Acconci façade at Storefront for Art and Architecture in New York in 1994. In advance of the show, I walked around the building with the receiver, taking location readings and charting the space. The path I walked was traced in dots of data, like digital breadcrumbs. During the show, the stationary antenna on the roof fed a monitor in the gallery with a real-time report on its location. The maps suggested that the gallery should become even more open than what the inside-outside dichotomy of the façade of the Storefront, with its hinged panels that expand the space of the gallery out onto the sidewalk, allowed, and that this could occur through emerging digital and social networks.

One year later, I reconfigured the project as *You Are Here: Museu* on the roof of Richard Meier's Museu d'Art Contemporani de Barcelona (MACBA) building. I had earlier given a lecture at the neighboring Centre de Cultura Contemporània de Barcelona (CCCB), which overlooks the roof of the MACBA. The roof is a very flat, abstract plane. Once I noticed that its skylight creates a surface like a digital display, square by square by square, the decision to use the GPS to spell the word M-U-S-E-U was clear. It happened then, on the spot.

GLOBAL POSITIONING

New York and Barcelona, 1994–1995 — These days, orienting yourself is becoming increasingly disorienting. Now, in order to answer that old question about where you are, it seems one has to leave the ground and travel into space, and more exactly into the cyberspace of a global satellite network. It is said that satellite positioning technology offers a definitive solution to this question, which some claim has troubled us from our origin: Where am I?

Where we are, these days, seems less a matter of fixed locations and stable reference points and more a matter of networks, which is to say of displacements and transfers, of nodes defined only by their relative positions in a shifting field. Even standing still, we operate at once in a number of overlapping and incommensurable networks, and so in a number of places — at once. Orienting oneself in this open and ongoing interaction appears all the more imperative and all the more impossible. "Where am I" in what? Where am I, where? In the global market, in the universe, in the family, in a corporate database, in some collective history, in the city or the desert, in the Internet, on the information superhighway?

With the Global Positioning System, it is said, a definite answer can finally be provided with a precision verging on one centimeter. "GPS really allows every square meter of the earth's surface to have a unique address," as Trimble's *GPS: A Guide to the Next Utility* puts it. "Everyone will have the ability to know exactly where they are, all the time."[1]

But the space or the architecture of the information system that wants to locate us once and for all in space has its own complexity, its own invisible relays and delays. The difficulty of charting the spaces that chart the spaces, of mapping the scaleless networks of the very system that promises finally to end our disorientation, demands redefining the points and lines and planes that build the map and lingering in their strange spaces and times. *You Are Here* is an attempt to begin mapping this emerging space of information using its own technologies. These are drawings with satellites, not to pinpoint a location, but to experience the drift and disorientation at work in any map or any architecture — especially the architecture of information.

POINT

The GPS is a network of twenty-four satellites and five ground stations designed to provide anyone carrying a portable satellite receiver with a highly specific determination of his or her location, anywhere, anytime, and in any weather. It promises that people and their vehicles will never get lost…missiles and bombs, as well as airplanes, will land exactly where they ought to…and a world of stationary objects, from telephone poles to wetlands to private homes, will be fixed once and for all in their proper places.

How to locate a point with GPS? Stand somewhere open to the sky with a GPS receiver for ten minutes and collect a stream of position readings [SCATTER]. Atmospheric interference, military scrambling, and the "multipathing" bounce of the signals in a built environment combine to represent that stationary position as a complex scattering of points. Download the data to a computer from the receiver, and then download from a local base station its (scattered) readings for the same time period. Correct your readings and reduce the drift [CORRECTION], average the points, and learn where you were…within a few meters. In the computer, the satellites draw the points for you, and as the readings become more precise, the points grow to fill the screen [POINT].

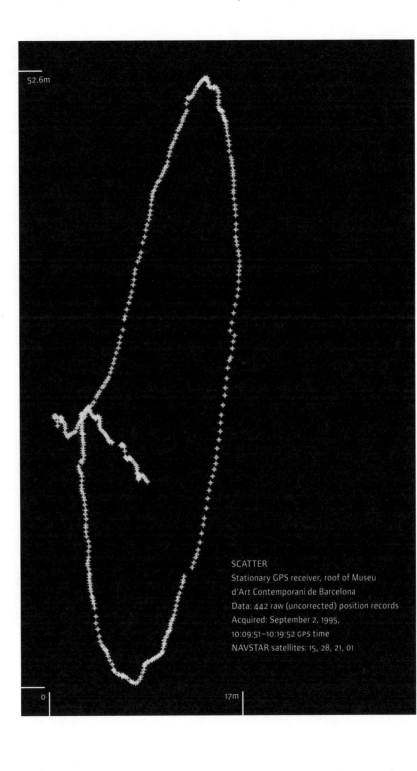

52.6m

SCATTER
Stationary GPS receiver, roof of Museu
d'Art Contemporani de Barcelona
Data: 442 raw (uncorrected) position records
Acquired: September 2, 1995,
10:09:51–10:19:52 GPS time
NAVSTAR satellites: 15, 28, 21, 01

0   17m

2.63m

0

13.88m

CORRECTION
Stationary GPS receiver, roof of Museu
d'Art Contemporani de Barcelona
Data: 442 position records, after differential
correction by reference to base station at
Tortosa, 40°49'15.118" N, 0°29'23.494" E

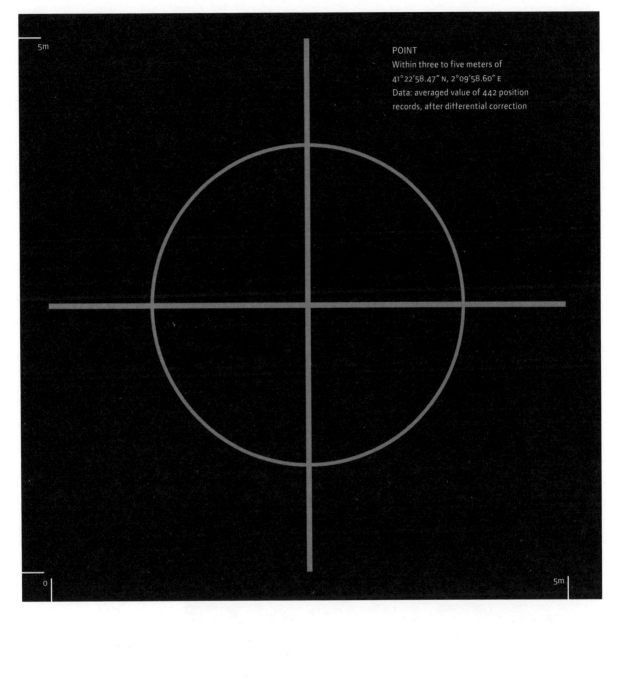

5m

POINT
Within three to five meters of
41°22'58.47" N, 2°09'58.60" E
Data: averaged value of 442 position
records, after differential correction

0

5m

5m

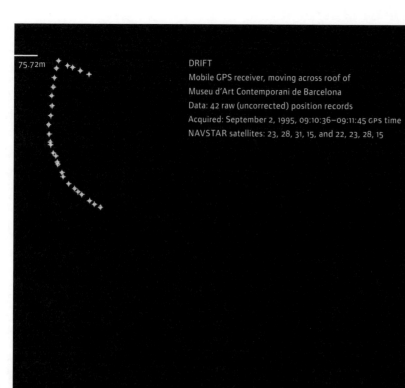

75.72m

DRIFT
Mobile GPS receiver, moving across roof of
Museu d'Art Contemporani de Barcelona
Data: 42 raw (uncorrected) position records
Acquired: September 2, 1995, 09:10:36–09:11:45 GPS time
NAVSTAR satellites: 23, 28, 31, 15, and 22, 23, 28, 15

0   31.57m

NAVSTAR Constellation:
23 28 31 15

A   : Sep 02 09:10:36 1995
41°22'58.56"N 2°09'59.15"E

B   : Sep 02 09:10:37 1995
41°22'58.54"N 2°09'59.10"E

C   : Sep 02 09:10:38 1995
41°22'58.53"N 2°09'59.10"E

D   : Sep 02 09:10:39 1995
41°22'58.52"N 2°09'59.10"E

E   : Sep 02 09:10:40 1995
41°22'58.52"N 2°09'59.15"E

F   : Sep 02 09:10:41 1995
41°22'58.52"N 2°09'59.11"E

G   : Sep 02 09:10:42 1995
41°22'58.53"N 2°09'59.12"E

H   : Sep 02 09:10:43 1995
41°22'58.54"N 2°09'59.10"E

I   : Sep 02 09:10:44 1995
41°22'58.55"N 2°09'59.09"E

J   : Sep 02 09:10:45 1995
41°22'58.55"N 2°09'59.07"E

K   : Sep 02 09:10:50 1995
41°22'58.54"N 2°09'59.08"E

L   : Sep 02 09:10:51 1995
41°22'58.54"N 2°09'59.07"E

M   : Sep 02 09:10:52 1995
41°22'58.55"N 2°09'59.06"E

N   : Sep 02 09:10:53 1995
41°22'58.56"N 2°09'59.05"E

NAVSTAR Constellation:
22 23 28 15

O   : Sep 02 09:10:55 1995
41°22'58.51"N 2°09'59.09"E

P   : Sep 02 09:10:57 1995
41°22'58.53"N 2°09'59.05"E

Q   : Sep 02 09:10:59 1995
41°22'58.55"N 2°09'59.02"E

R   : Sep 02 09:11:00 1995
41°22'58.58"N 2°09'58.97"E

S   : Sep 02 09:11:02 1995
41°22'58.59"N 2°09'58.95"E

T   : Sep 02 09:11:03 1995
41°22'58.61"N 2°09'58.92"E

U   : Sep 02 09:11:05 1995
41°22'58.63"N 2°09'58.88"E

V   : Sep 02 09:11:07 1995
41°22'58.66"N 2°09'58.84"E

W   : Sep 02 09:11:09 1995
41°22'58.68"N 2°09'58.82"E

X   : Sep 02 09:11:11 1995
41°22'58.71"N 2°09'58.78"E

Y   : Sep 02 09:11:12 1995
41°22'58.72"N 2°09'58.77"E

LINE

Take a walk, even a short one, with a GPS receiver. A minute and a half on the roof of the building leaves a faint collection of points [DRIFT], about one every two seconds, the oddly scattered remainder of a meeting with five satellites. Correct them differentially, and a line emerges [CORRECTION]. Moving is collecting points, which is to say, drawing. With a real-time display, you can watch yourself walking, charting, wandering…on the roof? On the screen?

The network is a machine for leaving traces, and so we can draw with satellites. The record of the interaction appears at the foot of each display: the identifying numbers of the NAVSTAR satellites, the time spent in contact with them, the number of data points collected by the receiver. What remains of that correspondence is something like a line, a sequence of points that registers the movement of the receiver across some physical space. But the line that results [LINE] from the transmission of data, charts more than one drifting pathway…across the roof, across a representation, across the screen. And in the network.

GPS location data, always a series of points, require that both movement (line) and stasis (point) be registered as drift in the zone of information, and so the map user operates in an oddly layered space, as if data and Earth were at once utterly independent of and somehow transparent to one another. The ostensible elements of architecture—points, lines, and surfaces—all find themselves transformed and redefined in the interactions of this network. This scaleless information zone constitutes not simply the representation of a preexisting space—as if built or physical space had some priority—but another space altogether. The possibilities of disorientation, not in the street or on the roof, but precisely in the database that promises orientation, are of an entirely different order, and GPS offers the chance to begin mapping some of these other highways as well: drift in the space of information.

18.87m

LINE
Mobile GPS receiver, moving across roof of Museu
d'Art Contemporani de Barcelona, from within
three to five meters of start point, 41°22'58.56" N,
2°09'59.15" E, to within three to five meters of
endpoint, 41°22'59.10" N, 2°09'58.52" E
Data: 42 position records, after differential
correction by reference to Tortosa base station

0

15.17m

IMPLIED PLANE

Over our heads, not like a canopy or a roof, but more like a panoply, the twenty-one active satellites and the three spares of the NAVSTAR constellation construct a strange space, in space. Guided by the five ground stations scattered around the globe, near the equator, their signals blanket the Earth, and their movements allow us to chart ours, at any time. Or rather, their incessant motion allows us to stabilize ours, because their motion is not only that of objects traveling in space, but that of constant broadcast, transmission, flow in the electromagnetic spectrum.

Any GPS position reading implies the interaction of at least three satellites (and the addition of a fourth allows an altitude calculation) and thus inscribes the active interface with an information network [CONSTELLATION]. Frozen in the second that defines a point's registration, the satellites constitute or imply a network of planes—not the planes that enclose a volume or a shelter, but the planes of a transmission, of the relay or passage of information at the speed of light and of information that amounts to nothing more than the record of their own positions [PLANES]. They move in a fourth dimension: they transmit a "pseudo-random code" for timing purposes, based on data from on-board atomic clocks, along with a message about their exact orbital location and the status of the system, in a few hundredths of a second. Their orbital paths define the movements that link them as a network, a series of nodes defined only in relation to one another, but the pathways followed by their radio transmissions are as much temporal as spatial. Constant real-time transmission grants them a certain ubiquity: invisible in their motion, they render everything visible. Without watching or listening, without the eyes and ears with which we figure surveillance, they nevertheless lay a grid over the totality of the Earth's surface; they define it as a totality and mark every position on that grid with a real-time address, a unique and singular geotemporal code. They make up an "information" or "orbital front," as Paul Virilio has called it, that operates in a fourth dimension, an "exo-spheric" and strictly temporal dimension, "that of the real time of ubiquity and instantaneity…less physical than micro-physical."[2] Twenty thousand kilometers and six one-hundredths of a second overhead, they are transmitting—now.

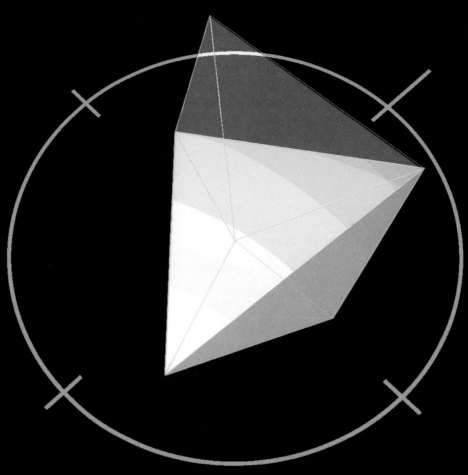

CONSTELLATION
NAVSTAR satellites: 23, 28, 21, 15, and 04, 07, 14, 28
over Barcelona, September 2, 1995, 09:10:36 GPS time,
and September 1, 1995, 15:28:37 GPS time
X, Y axis (azimuth): position of satellite away from north
Z axis (altitude): position of satellite above horizon

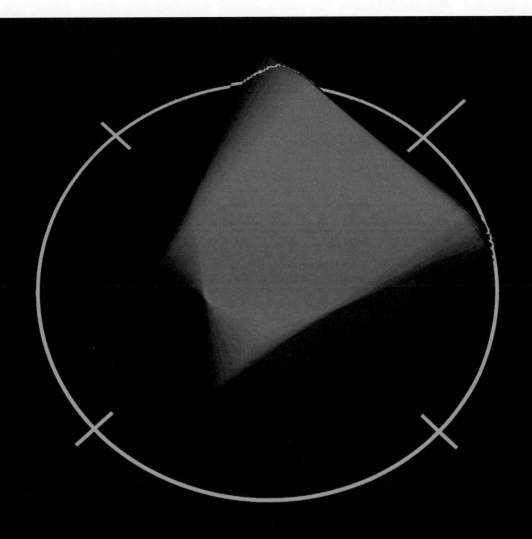

PLANES
NAVSTAR satellites: 15, 28, 21, 01, 23, 22, 31,
over Barcelona, September 2, 1995,
09:37:17–10:30:14 GPS time

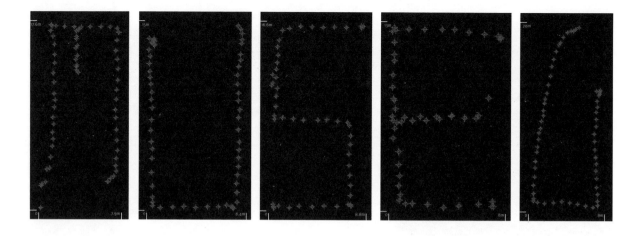

LETTERS

Drawing with satellites…from points to lines, the GPS network defines physical space—whether the roof of a building, the public space of a city plaza, or the wide open sea—as a surface for inscription. Not just for the data that pin down the point or identify the place with an "address" precise enough to call for help or to direct a cruise missile, but for writing in the narrow, everyday sense. Take a walk with a receiver, pace out the steps in the form of letters, and watch the display unfold on the computer screen: writing with data points. Lines make letters, and letters join into words: information of another order, a word written—and read—not from above, not even seen in any conventional sense, but traced on a purely digital surface.

Walking, under the satellite sky, is writing, somewhere else. On the roof of the MACBA, the carefully rational grid that frames the skylights marks out the space of a virtual word in Catalan: M U S E U. The structure of the word is built into the building, like the characters latent in the LED display of a digital clock: the points await their configuration into any number of signifying combinations. Dot by dot, the GPS data points are transformed into the word that names the building and the institution. Now, at once fleetingly and forever, the roof reads "museu," museum…a machine for collecting and preserving, for inscribing, the traces of a culture. Two styles or spaces of inscription cross with each other, two conventions of description or interpretation: the precise coordinates, in longitude, latitude, and altitude, of the building, and the letters of the name, both common and incipiently proper, of the institution which, as such, sets off and demarcates that space as a zone of preservation and display. Two information networks, two virtual spaces, condensed into a map of this *museu*…and exposed.

MAP

How to compile a map with GPS? *You Are Here* maps the spaces of a building, the Museu d'Art Contemporani de Barcelona, and then installs that map in and on the building itself: not in the name of self-reference, but rather of superimposition, of the overlay of asymmetrical spaces. Build up a series of successive point and line measurements over two days in September: I stood ten minutes apiece for five points [FIVE POINTS], walked two lines on the roof [ALIGNMENT, FACADE], and constructed a set of five letters. The data and drawings—on the wall, on the monitors, on the building, and on these pages—are the traces of an interaction with the satellite network, and the physical space is layered over and folded with the immaterial remnants of this encounter. The passage of data through the electromagnetic spectrum and cyberspace leaves its mark on the site of reception—not with the destructive force of an explosion, but with the silent insistence of images, light, and writing.

The composite map is a series of layers, corrected and averaged points traced over one another in the memory of a computer. The layered data are correlated by reference to a quasi-arbitrary point: the so-called 0/0 reference point enables the digitized data to be coordinated with the space of their reference. On the composite map [BUILDING], not all the points recorded by the GPS receiver—even the averaged points—fit into the space defined by the walls of the museum. And even with the most accurate receiver available on the market and the most precise corrections possible, the point is always divisible into a series of points somewhere in the zone of an expected point. The GPS information refers to, but does not simply represent, the space it maps: it exceeds, transforms, and reorganizes that space into another space. Not a representation of a space, but a space itself…or rather, spacing itself, passage and inscription, light and motion, transmission and interface. GPS can locate a target to within a few meters, measure the movement of a mountain after an earthquake, keep an airplane on course, direct a 911 response team to your doorstep—and this active intervention obliges us to take these maps and readouts seriously, obliges us to think of these computerized maps as real spaces, at least as real as anything else (the building, for example). Perhaps there is more than one dominant definition of this, or any, space. The composite map, in its compilation and complication, charts a digital ground, a space of digital points—a space in which we think and act and move, every day.

24.53m

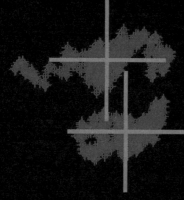

FIVE POINTS
Stationary GPS receiver on roof of Museu
d'Art Contemporani de Barcelona
Data: averaged values of 1828 position readings,
after differential correction
Acquired: September 2, 1995, in five separate
sequences, 09:37:17–10:30:14 GPS time
NAVSTAR satellites: 15, 28, 21, 01, 23, 22, 31

1: 41°22′58.93″ N, 2°09′58.25″ E
2: 41°22′58.75″ N, 2°09′58.35″ E
3: 41°22′58.56″ N, 2°09′58.56″ E
4: 41°22′58.47″ N, 2°09′58.60″ E
5: 41°22′58.30″ N, 2°09′58.87″ E

0

20.52m

27.63m

ALIGNMENT
Five points and line, stationary and
mobile GPS receiver on roof of above
Museu d'Art Contemporani de Barcelona
Data: 1870 position records, after
differential correction
Acquired: September 2, 1995, in six separate
sequences, 09:10:36–10:30:14 GPS time
NAVSTAR satellites: 15, 28, 21, 01, 23, 22, 31

0

23.61m

78.61m

FACADE
Five points and two lines, stationary
and mobile GPS receiver on roof of
Museu d'Art Contemporani de Barcelona
Data: 2010 position records, after
differential correction
Acquired: September 2, 1995, in seven sepa-
rate sequences, 09:10:36–10:30:14 GPS time
NAVSTAR satellites: 15, 28, 21, 01, 23, 22, 31

0

86.71m

92.68m

BUILDING
Nine points and two lines, stationary
and mobile GPS receiver on roof of
Museu d'Art Contemporani de Barcelona
Data: 2784 position records, after
differential correction
Acquired: September 1, 1995, in four separate
sequences, 10:07:03–10:50:44 GPS time,
and September 2, 1995, in seven separate
sequences, 09:10:36–10:30:14 GPS time
NAVSTAR satellites: 15, 28, 21, 01, 23, 22, 31, 14

0

88.55m

MUSEU

What happens to the museum in an age of digital mapping, of real-time data flows and virtual realities? The Museu d'Art Contemporani de Barcelona opened its doors in November 1995, but the building itself is designed to open permanently to the outside, flooded as it is by the natural light that enters through the glass of the roof and the façade. "It's the Mediterranean light that makes this building unique," says the architect, and the light symbolizes a more general openness: "Today a museum is more than a container for works of art," he says, "it's a place where people come together, a social place as well as a place for contemplation."[3] But what is a "social place," now that information moves through buildings and places at the speed of some other, unnatural light? New public spaces and new modalities of being together or at odds emerge in the vectors of data flows, in the media and online, places invisible to the eye of the rational-critical thinker. And what becomes of the aura of the building and of the artwork contained and contemplated in it, now that uniqueness flickers in the light of the monitor, fiber optics, and satellite relays? Could we ever really imagine an inside that remains pure, a space of consolidation and identity that resists the intrusion of everything foreign, secured by the frontier of a line? The barriers between public and private, outside and inside, always questionable, have long since been eroded and transformed by so many complications, of which digital flows are only one figure. The museum is a node, fixed and unfixed by its changing position in different networks, and try as we might, we can never simply be inside or outside its space. Without these reliable boundaries, at the museum or anywhere, disorientation becomes less a problem to be solved than an irreducible condition of possibility of our movements in space and time.

In the fall of 1995, MACBA became both the subject of and the surface on which to register the flows and displays of a GPS digital mapping network. *You Are Here: Museu* installed a real-time feed of GPS satellite positioning data from an antenna located on the roof of the gallery and displayed in it, together with the record of mapping data collected in September, in light boxes and inscribed on the walls and floors of the gallery. These interferences between the digital and built space are experienced as drift: the map as the impossible alignment of the museum building with the electronic space engaged in mapping it.

92.68m

YOU ARE HERE: MUSEU
Nine points, two lines, and five letters,
stationary and mobile GPS receiver on roof
above Museu d'Art Contemporani de Barcelona
Data: 3079 position records, after
differential correction
Acquired: September 1, 1995, in nine separate
sequences, 09:48:13–15:30:13 GPS time,
and September 2, 1995, in seven separate
sequences, 09:10:36–10:30:14 GPS time
NAVSTAR satellites: 04, 07, 14, 18, 15, 28, 21,
01, 23, 22, 31

0

88.55m

FRONTS

In 1993, the Storefront was given a new "front," a permeable membrane of pivoting doors aimed at blurring the distinction between the street and the gallery. The architect says: "this project has two extremes; totally closed, and totally open.... If it is closed, it is a wall with lines on it. When it is open, the outside is inside and the inside is outside."[4] In its symmetry, this position leaves room for a very limited interpretation of both outside and inside. Like the opposition between open and closed, the distinction between inside and outside stays firmly in place—the wall with lines on it remains itself a line. But can we imagine an inside that remains pure, that resists the intrusion of everything else, secured by the frontier of a line? Storefront is a node, defined and redefined only by its changing position in different networks, and try as we might, we could never be simply inside or outside the space. One cannot choose to open or close oneself to the outside, as one chooses to open a door or a wall. But without these reliable boundaries, at the Storefront or anywhere, disorientation becomes less a problem to be solved than an irreducible condition of possibility for our movements in space and time.

In March and April 1994, *You Are Here: Information Drift* closed the doors of this new façade in order to open it onto and inscribe it into a usually invisible network, an orbital or digital front. Turned into a satellite receiver, the Storefront became both the subject of and the surface on which to register and display the flow of digital mapping with GPS.

50 MINUTES, 5 POINTS (STOREFRONT)
Five points, stationary GPS receiver on roof
of Storefront for Art and Architecture
Data: 1349 position records, after differential
correction by reference to Trenton Base Station,
40°13' 14.014" N, 74°45' 24.640" W
Acquired: January 25, 1994, in five separate
sequences, 16:41:06–16:51:06, 17:08:21–
17:17:59, 17:19:05–17:29:06, 17:31:28–17:41:28,
16:56:13–17:05:53 GPS time
NAVSTAR satellites: 01, 05, 12, 15, 20, 21, 23, 25

90'

0

125'

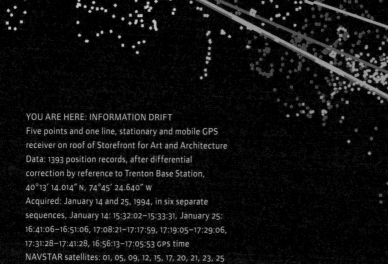

117'

0

125'

YOU ARE HERE: INFORMATION DRIFT
Five points and one line, stationary and mobile GPS
receiver on roof of Storefront for Art and Architecture
Data: 1393 position records, after differential
correction by reference to Trenton Base Station,
40°13′ 14.014″ N, 74°45′ 24.640″ W
Acquired: January 14 and 25, 1994, in six separate
sequences, January 14: 15:32:02–15:33:31, January 25:
16:41:06–16:51:06, 17:08:21–17:17:59, 17:19:05–17:29:06,
17:31:28–17:41:28, 16:56:13–17:05:53 GPS time
NAVSTAR satellites: 01, 05, 09, 12, 15, 17, 20, 21, 23, 25

PIXELS

"We need to know for certain where we are," says a man holding a map titled *Los Angeles Fires and Civil Unrest*, in a corporate brochure for real-time GPS mapping.[5] The open reference of the statement summarizes the promise and the dream of GPS: accurate positions, instantaneously and continuously. One newspaper report on GPS in passenger cars was headlined: "In Japan, They May Never Ask for Directions Again."[6] Not simply for pilots and engineers and ambulances, but for everyone, anyone, facing a location crisis. "With today's integrated circuit technology," suggests one manufacturer's handbook, "GPS receivers are fast becoming small enough and cheap enough to be carried by just about anyone. That means that everyone will have the ability to know exactly where they are, all the time. Finally, one of man's basic needs will be fulfilled.... Knowing where you are is so basic to life, GPS could become the next utility."[7]

Another announcement for a GPS image-mapping software package, combining GPS, GIS, and remotely sensed images from Landsat and SPOT, promises that it can finally deliver a reliable answer to the questions that continue to vex even the users of the most powerful maps: "'Which pixel am I standing on?' or worse, 'Where am I?'"[8] Not "Where am I?" on the Earth, but where on the map? At a time when these digital technologies seem to offer great leaps in our ability to locate ourselves and when not only frightened urbanites, but some of our most radical social critics, are calling for "an aesthetic of cognitive mapping," a critical analysis of new mapping technologies seems imperative.[9] But perhaps the sense of what's "worse" conveyed by the GPS announcement needs to be rethought: the older and perennial question of "Where am I?"—the question that gives rise both to panic and to new discoveries—has been replaced or displaced by a still stranger interrogative, "Which pixel am I standing on?" How to orient oneself in the cyberspace that promises orientation? What could it mean to stand on a pixel? Who or what stands in or on the data space of a pixel? The orbital front at once offers unprecedented mapping and positioning powers—capacities which for better or for worse should not be underestimated—and opens new questions that challenge the most basic ways we think about space.

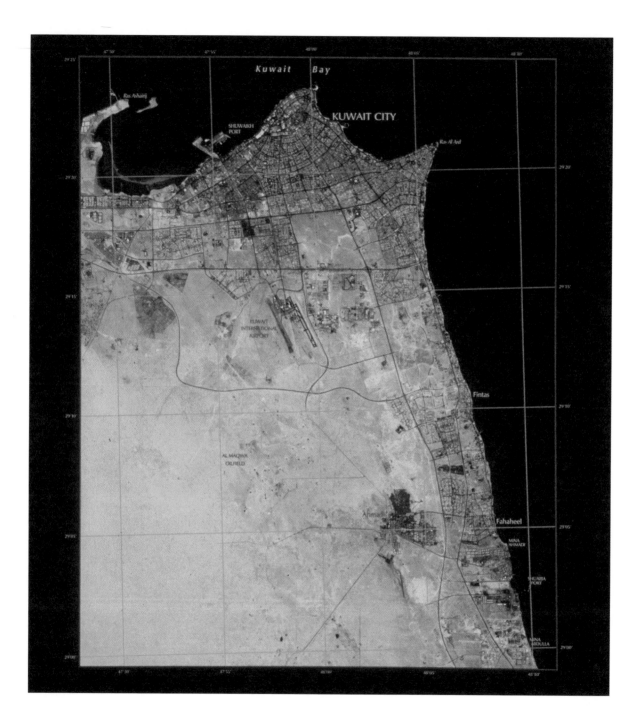

From the Kuwaiti National Database, Intergraph Corp.

2 Kuwait: Image Mapping

From within the spaces of the incriminated technologies themselves

New York, 2012 — At about the same time that GPS was emerging, if not as a household word or appliance, at least as a topic of public discussion, something similar was happening with digital maps. And it was because of the Gulf War—not only the war as a military fact, as a matter of the deployment of lethal violence and destruction, but the war as a public and a media event, as a matter of images.

I had seen an advertisement for Intergraph software just after the end of the war in early 1991. It was selling digital cartography tools, and condensed into the script was somehow the entire narrative of the war. A dispute about borders, a spectacularly expensive project in total mapping by the sovereign of a small authoritarian emirate, a destructive war that mobilized digital-spatial-imaging weaponry (and the media) as never before, and an aftermath that was presented as fundamentally continuous—in software terms—with what had preceded it. The ad featured nothing less than a satellite image: an image of destruction, promising reconstruction and signifying a database. "The Kuwaiti government used an Intergraph system to digitally map its country. The database now provides the foundation for rebuilding Kuwait's infrastructure."[10]

I found it startling: on the one hand, the satellite images of burning oil fields had become an icon of wanton destruction by the Iraqi state and offered an ex post facto rationale for the war, but in combination with a GIS database that plotted every tree, building, and other landmarks in Kuwait, it also became a preservation image. It was used to reconstruct the ruined postwar city.

At the time, it was said, space was being revolutionized by the technologies of digital communication, imaging, and cartography. There was a break between events that were occurring in space and their virtual descriptions, and this carried into intellectual debates around the war. There were of course many enthusiasts. Opponents of the war, too, took the claims about revolution seriously—or at least literally—and more or less inaugurated an entire genre of technocritical discourse about war after the end of the Cold War. Jean Baudrillard argued in a series of newspaper essays—published before, during, and after the fighting—that the Gulf War would not take place, was not taking place, and had not taken place, which is to say that the notions of space and place, and with them what it means for something to happen, had undergone a profound and debilitating transformation.[11] In his book on the Gulf War, Paul Virilio suggested that real-time television had rendered democracy obsolete.[12] Neil Smith claimed that "GIS and related technologies" had contributed to the loss of some two hundred thousand lives in "the killing fields of the Iraqi desert."[13]

Ordering a series of Landsat images and reproducing them along with images from the Kuwaiti Municipal Database in *Documents Magazine* was an attempt to engage some of these overlapping spatial and political questions, and to do it from within the spaces of the incriminated technologies themselves. Military inventions will always be designed to interpret and conquer space in new ways. The challenge, I felt, was to harness these devices for other ends, all the while underlining just these military origins.

KUWAIT: IMAGE MAPPING

New York, 1992 — Underneath an image of a computer screen displaying a map of Kuwait City, a magazine advertisement from 1991 reads: "Three days after the cease-fire, the U.S. Army Corps of Engineers was in Kuwait, rapidly assessing damages. Using thousands of computerized maps compiled on an Intergraph system. Because all maps were survey-entered and field-verified, the Kuwait Municipal Database is one of the most accurate and comprehensive in the world. Never before has such data been available for planning, recovery and reconstruction…. If you are a contractor planning to do work in Kuwait, find out how the Kuwait Municipal Database can help you."[14]

Vector-based and field-verified layers of data, from Kuwaiti National Database, Intergraph Corp., 1991.

In other words, already before the conclusion of the first Gulf War, the data and maps that guided the reconstruction of Kuwait were already available. On another Intergraph poster titled: "Kuwait City: Image Mapping...the Integration of Remote Sensing, GIS and Digital Cartography," the map is explained as composed of two satellite images, taken on July 17, 1987, and August 31, 1990.[15] These images are shown manipulated and superimposed onto what is called the "Kuwait National Database," eight years in the making and completed weeks before the Iraqi invasion, which correlated the digitized satellite (raster) data with other mapping (vector) data, thereby converting it into a map: just in time for the space it represented to be destroyed by the invading Iraqi army.

The interfaces here are multiple in that this map superimposes data systems and imaging systems. It keeps the viewer overhead—at hundreds of miles above the Earth, in two satellites—and on the ground—with tape measures, spread across years of imaging and data collection—and at the same time in the archive of a database.

According to a brief account in *Armed Forces Journal International*, the government of Kuwait gave Intergraph permission to distribute the database "to firms that receive contracts to help rebuild the Mideast nation. The database includes the location and shape of 145,000 buildings in Kuwait City and complete records of all roads, bridges, paved areas, and parking lots, as well as all utility, telephone, and power lines. It even contains the precise location of all of the trees within Kuwait City prior to Operation Desert Storm. The database, completed by the Japanese firm Mitsui in June 1990, took eight years to complete and cost the Kuwaiti government $30-million."[16]

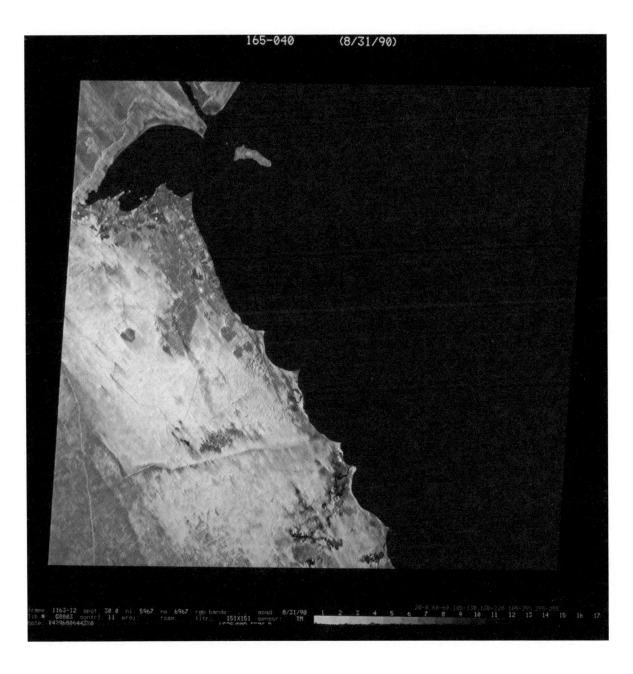

This image, taken at the beginning of the Gulf War on August 31, 1990, was used as the color base layer of the Kuwaiti National Database.

IMAGE: LANDSAT, FROM EROS DATA CENTER

The result: an image that was designed for the purposes of a total planning and surveillance regime, then used to plan the U.S. counterinvasion of Kuwait, and then reused as a preservation and reconstruction blueprint. In principle, every uprooted tree and every occupied or destroyed building could be restored according to the data in the image map.

An image is not necessarily a map. The Intergraph image of Kuwait City is an example of "image mapping," a technical term that describes the overlay and integration of remote-sensing imagery with a geospatial database. According to Intergraph, the "Kuwait City image was created with data collected by two commercial satellite systems, Landsat and SPOT.... Spectral (color) information of the 30-meter ground resolution Landsat data was digitally combined with the spatial (textural) information of the 10-meter resolution of SPOT panchromatic data." The new composite image is treated as a georeferenced grid of pixels — picture elements. Onto the pixels are superimposed vector files, scaled lines generated on an *x-y* coordinate system that contains the lines, symbols, and labels of conventional maps. "Cartographic annotations, symbology, and marginalia were added in vector form using the raster image as a co-registered backdrop. Features and labels were extracted from a GIS database originally created for analysis and resource management."[17] This gives the image a scale and orientation and allows it to be coordinated with information in city and military databases or with other maps. The layers of the image map can become visible or invisible to reveal or conceal transportation, communications, power, or water and sewer network maps, for instance.

And so it was not a big jump from "resource management" to war fighting. Interviewed in a promotional Intergraph publication just after the war, one company executive traced a direct line from "one of the most complete GIS projects ever developed" to the tactical conduct of the war. He noted that the Kuwaiti National Database, nicknamed "Kudams," contained "13.8 million features representing topographic, property, and utility data, in a remarkable level of detail" and underlined that "Kudams provided vital intelligence information throughout the time Kuwait was occupied. The system was instrumental in putting together operational plans for liberating the country."[18]

At the start of the post–Cold War, had any 5,800 square miles been more intensively mapped than these? These maps and mapping programs facilitated governing the country, delimiting its uncertain borders, fighting a war in and over it — and starting all over again. By the early morning of January 17, 1991, when the coalition air attack on Iraq and occupied Kuwait began, more than one line had already been drawn in these sands. The Gulf War was a conflict over just where and how to draw them.

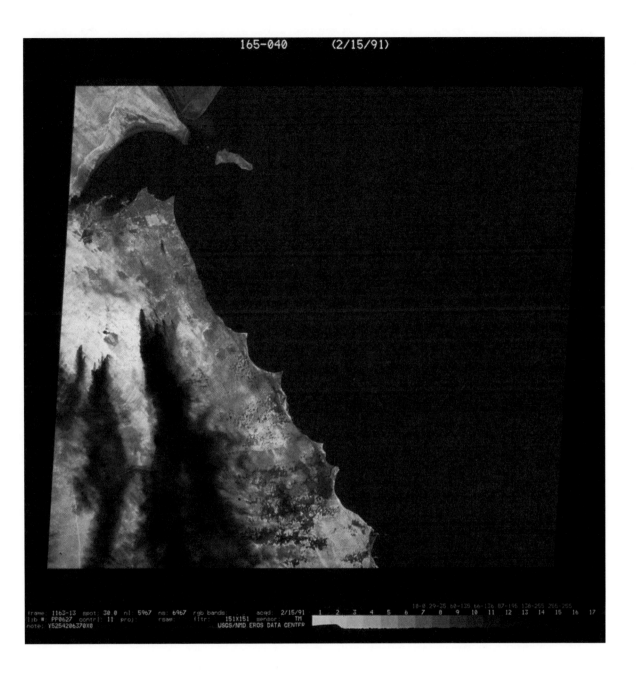

In this image, taken during the Gulf War on
February 15, 1991, smoke from oil fires obscures
much of the landscape.

IMAGE: LANDSAT, FROM EROS DATA CENTER

A *Washington Post* reporter even referred to the invasion of Kuwait itself as a "failed exercise in cartography-by-tank." Writing from the "steamy port town" of Umm Qasr, Caryle Murphy noted that the border was near it, "but because the frontier was never marked, few people knew exactly where the line lay. Iraqi leader Saddam Hussein hoped to clear up this ambiguity by invading and annexing Kuwait."[19]

He failed, and maps were not only at the source of the dispute, but also instrumentalized in its conduct. The Gulf War was a battle unprecedented in its reliance on maps, from the digital ones stored in the on-board memories of cruise missiles to the commercial satellite data purchased by the Pentagon during the war. Paradoxically, the detailed maps commissioned years before prefigured both the wartime and postwar need for their extensive detail and the erasure of the lines they interpreted: as if the possibility of this erasure, as if the war itself, were already inscribed in the drawing of the map.

It's easy now to look at Kuwait from 561 miles above. These three images were made from remote-sensing data gathered once every three days along the Landsat satellite's track around the Earth. Images like this were produced before, during, and after the war over Kuwait. Although it was widely reported that satellite surveillance provided much information to Allied forces during the war, their military imaging capacity was severely limited by weather, night, and other obstacles, such as smoke. As one headline in the *Los Angeles Times* put it, "U.S. reconnaissance satellites aren't all-seeing, so don't expect miracles."[20]

For these and other reasons, the military made extensive use of available images from SPOT and Landsat satellite images for maps and targeting. "Satellite-based map information of Kuwait and Saudi Arabia was lacking as the Desert Shield buildup began in August, 1990. The Defense Mapping Agency and other organizations made extensive procurements of French Spot and US Eosat/Landsat data to update maps as the crisis escalated."[21] The U.S. Air Force obtained more than a hundred SPOT scenes of the region, including many of central Baghdad, which went into battle overlaid on pilots' digital terrain maps. "It was absolutely a life-saver," an officer told a journalist later. "It provided a never-before-seen capability in the field of mission planning."[22]

The Gulf War famously came to an end with a chaotic, map-driven slaughter and the images that resulted. On the night of February 25, 1991, flying in a Joint Surveillance Target Attack Radar System (J-STARS) Boeing 707, analysts were mapping what has since come to be known as "the highway of death." J-STARS radar technology detects moving ground forces on a battlefield, superimposes those tracks of movement on the ground onto stored maps, and distributes them to attackers in "near real time." That night, the J-STARS radar image mapped a

In this image, taken on February 23, 1991, smoke
from oil fires has spread widely across the terrain.

IMAGE: LANDSAT, FROM EROS DATA CENTER

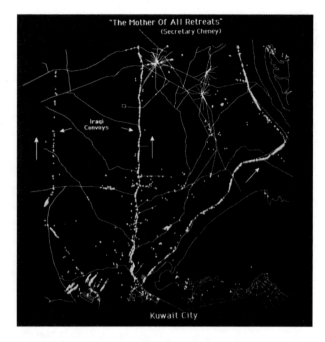

This J-STARS map, showing Iraqi convoys leaving Kuwait City, is titled and annotated by Grumman Corporation. PHOTO: GRUMMAN CORPORATION

retreating Iraqi convoy stretching two miles along the only six-lane highway north from Kuwait City across the border to Basra, and U.S. pilots destroyed it at the Al-Mutlah Ridge.[23]

According to General H. Norman Schwarzkopf's *It Doesn't Take a Hero*, other images of that highway, taken from the ground, probably ended the war. "Journalists were now interviewing…pilots who'd been hitting the convoys fleeing Kuwait [and]…reporters who had once been part of the media pools had taken pictures of Highway 6, where we'd bombed a convoy Monday night…. Washington was ready to overreact, as usual, to the slightest ripple in public opinion. I thought, but didn't say, that the best thing the White House could do would be to turn off the damned TV in the situation room."[24]

On this battlefield, the images were in combat and in turmoil.

And other satellites, not just U.S. and French, were also taking pictures over Iraq and Kuwait. "Soviet satellite coverage over Kuwait resumed on February 7th," reported *Armed Forces Journal International*, "when Kosmos 2124 picked up where 2108 left off. The replacement satellite was in an ideal position to gather 'convincing' photos on February 22nd—the day before [Iraqi Deputy Prime Minister Tariq] Aziz's visit to Moscow. And, unlike its predecessors orbiting over Kuwait, Kosmos 2124 can return packages to earth while remaining aloft. We may never know

"Highway of Death" in Northern Kuwait, March 12, 1991.   © PETER TURNLEY/CORBIS

whether the Soviets actually showed the Iraqis satellite imagery, or whether such photos played a role in convincing Saddam to end the lopsided conflict. If that did happen, it would be the first time satellite photos halted a major conflict."[25]

Whether or not this is true, the question itself prefigures what would become the central political, rather than military, role of satellite imagery in the following decade—as *evidence*, presented both to leaders and to civilians, in debates about the conduct of war.

An eighty-by-sixty-foot photomosaic of Africa,
assembled from imagery taken on the Argon KH-5
satellite missions (part of the Corona project)
between 1961 and 1964. IMAGE: U.S. GEOLOGICAL
SURVEY AND KEITH CLARKE, PROJECT CORONA AT
THE UNIVERSITY OF CALIFORNIA AT SANTA BARBARA

3   Cape Town, South Africa, 1968:
Search or Surveillance?

The hinterlands of the Cold War—
also of interest to the Corona cameras

New York, 2012—President Bill Clinton's Executive Order 12951 in 1995 did more than release thousands of spools of Cold War–era high-resolution satellite imagery to the public.[26] Although ostensibly aimed at opening up and clarifying the past, it actually pointed toward and even enabled a future. The vivid, if sometimes blurry resolution of those spy images taken in the 1960s and 1970s made us wonder how close the military could see *now*.

Compared with the highest-resolution imagery that was then available (Landsat, 30 meters; SPOT, 10 meters), the Corona images, which were the heart of the declassified archive, offered an unprecedented opportunity to look close up at a distance. But despite the *glasnost* in Washington, a complex cultural, political, and legal cloud hung over contemporary satellite imagery in the mid-1990s. While the International Criminal Tribunal for the former Yugoslavia debated whether or not to admit images of the Srebrenica massacre as evidence of war crimes, *Wired* effusively announced the soon-to-launch next wave of what would become the first commercially available imagery that could be purchased from high-resolution satellites, complete with lavish (and simulated) "images" of Washington, D.C., and Paris.[27]

Since the commercial high-resolution satellites had not yet been launched, I decided to look back. Corona satellites were always tasked to certain locations, so the images in the archive were heavily weighted toward the predictable ones of the early 1960s and 1970s—China, the Soviet Union, and Eastern Bloc countries. But the hinterlands of the Cold War, the spaces where proxies battled in smaller but more lethal wars, were also of interest to the Corona cameras. And so, there were myriad images from which to choose.

According to Keith Clarke, the Argon satellite missions (part of the Corona system) were used to construct the "first high resolution image of Africa," an eighty-by-sixty-foot photomosaic.[28] This set me on a trail of looking for a Corona image of Cape Town, South Africa, a place I know well. I grew up there, so I could be a specialist in the art of image interpretation. However, there were only a few high-resolution KH-4B satellite images taken over Africa—a few in Egypt and a few, rather coincidentally for my purposes, over Cape Town. Rather than speculate on what the U.S. government sought in Cape Town in 1968, I chose to highlight the quality of the images and what they could tell us about transformations on the ground, physical and political, in South Africa.

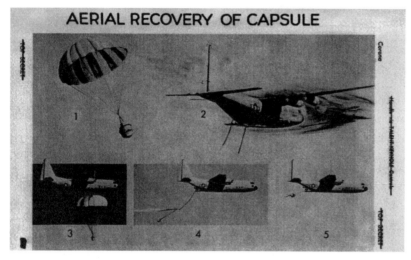

"The planned recovery sequence involved a series of maneuvers, each of which had to be executed to near-perfection or recovery would fail." DIAGRAM FROM KENNETH E. GREER, "CORONA," *STUDIES IN INTELLIGENCE*, SUPPLEMENT 17 (SPRING 1973), PP. 10–11

KH-4A Corona imagery, mission 1049-1, taken on December 17, 1968. Although the imagery has been declassified, the TOP SECRET RUFF marking is still visible on this roll of film.

SEARCH OR SURVEILLANCE:
WHAT CAN WE DO WITH WHAT WE SEE THERE NOW?

Boston, 1998 — "Space reconnaissance is traditionally divided into categories," we learn in the "Corona Program Profile" published by Lockheed Martin after the program's existence had been revealed in 1995. "One is called 'Search,' and is dedicated to answering the question, 'Is there something there?' Corona was designed to photograph large contiguous areas in a single frame of film in order to answer that question. A second observation function is 'Surveillance.' Surveillance is required after one has decided that 'There is something of interest there,' and says 'I want to continue to watch that something, learn more about it, identify it and classify it.' In most cases, bona fide surveillance was beyond Corona's capability."[29]

An American KH-4B satellite passed over the southwestern tip of Africa on November 11, 1968, leaving a trail of imagery behind for us to examine. Hundreds of miles above Cape Town, it exposed something political, opening up a landscape of data and of history in the image. Search, or surveillance? What can we do with what we see there now?

The privilege of seeing closely from great distances has until very recently been reserved for governments, spies, and militaries. It was only with President Clinton's 1995 release of, as they were described, "certain scientifically or environmentally useful imagery acquired by space-based national intelligence reconnaissance systems" that examples of high-resolution imagery became easily available. Most of the images declassified with that order were from the so-called Corona missions, the first American reconnaissance satellites, which orbited the earth on "top secret" missions from 1960 through 1972. Now hundreds of thousands of these photographs from space are in the public domain, many providing detailed imagery at a ground resolution of 5 to 7 feet, or 2 meters.[30] Compared with the best satellite images previously available to the public, from Landsat and SPOT, it seems that we—or at least some of us—are getting closer and closer from farther and farther away.

Corona's images, it is said, were designed for searching, not for surveillance. Today, the distinction between search and surveillance has become somewhat less sharp. To inquire about the existence of something and to investigate and watch over it can now happen simultaneously and from enormous distances in striking detail. Increasing the resolution implies erasing the distinction between existence and identity—"high resolution" means that looking for things and looking after them, searching and "bona fide surveillance," can increasingly take place in the same gesture.

Military satellites now regularly download digital imagery in the range of 50-centimeter to even 10-centimeter resolution to a few privileged eyes, and

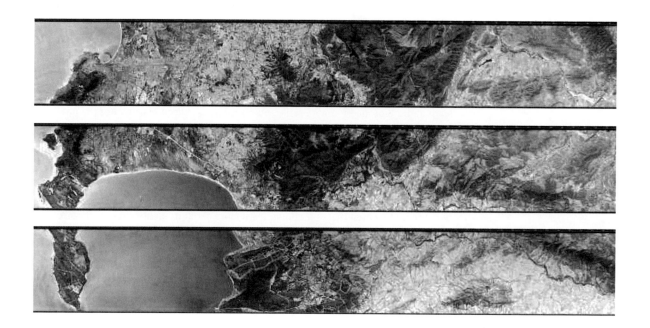

commercial enterprises promise virtually real-time delivery of 1-meter imagery from a host of new private satellites to anyone who can afford them. The declassified Corona archive makes older pictures widely available, and already what were Russian military spy satellites are now providing fresh 2-meter resolution imagery on the open market. Some 90 percent of the globe will soon be routinely monitored in a 1-meter grid — to monitor, learn about, and identify at the same time.

Full-frame images from KH-4B Corona satellite, Cape Town, South Africa, November 1968.

In November 1968, the Corona satellite passed over Cape Town. What remains for us to see in these blurry, high-resolution pictures? We don't just see things on the ground, places or people in and around Cape Town, apartheid city, seat of the South African Parliament today as then. What's there is data, like it or not, and now we can look at and into the images to monitor in our turn the creation of this new datascape. Data need to be interpreted and never can be, fully. Scanning the surface of the image, scale can disappear, while shapes and textures, differences and identities, threats and possibilities, statements and metaphors emerge, moving in and out of the contexts that the long strips of film provide, but never guarantee. Things become unrecognizable here, familiar features decompose as others come sharply into focus. Today, we can search, and watch, across many degrees of magnification, for the future in this image.

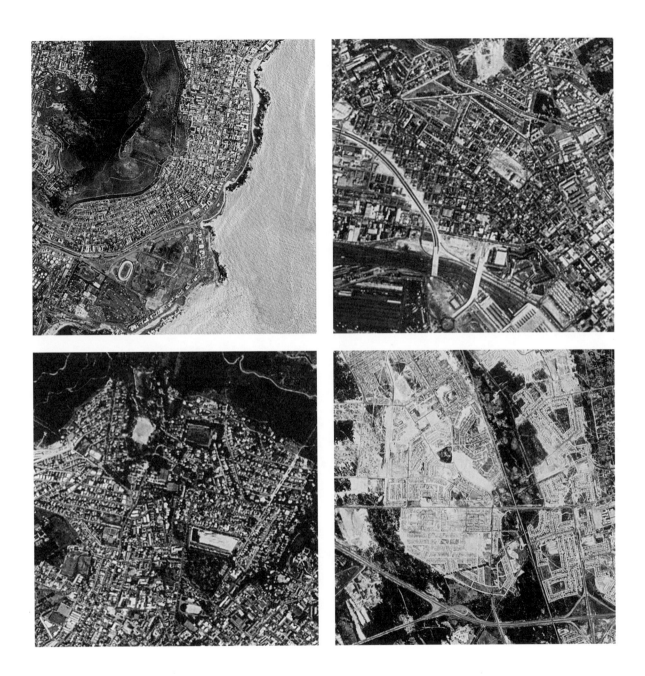

Details from KH-4B imagery, Cape Town, South
Africa, November 1968. Top left: Signal Hill;
top right: District Six; bottom left: Oranjezicht;
bottom right: Gugulethu.

Details from KH-4B imagery, Cape Town, South
Africa, November 1968. Top: nine ships off
Cape Town harbor; bottom: Seal Island, False Bay,
south of Mitchells Plain and Khayetlitsha.

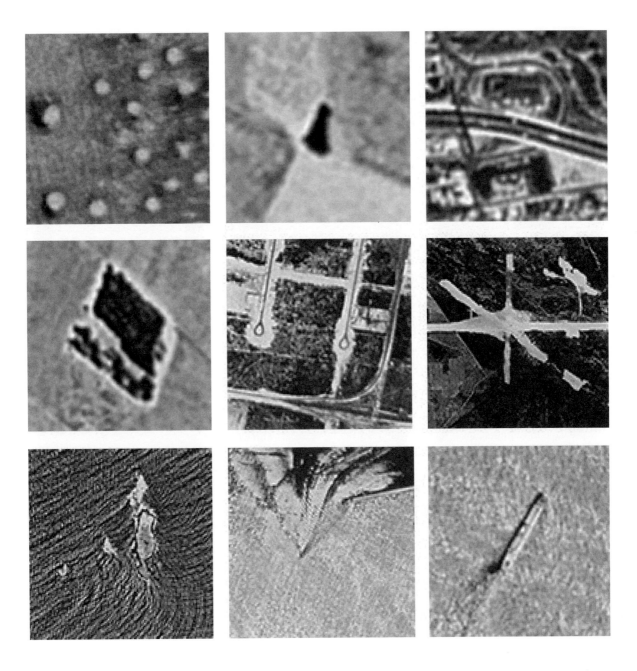

Details from KH-4B imagery, Cape Town, South Africa, November 1968. Top row, left to right: trees, path between two vineyards, cars on De Waal Drive. Middle row, left to right: unidentifiable object, dead-end roads, N2 intersection under construction. Bottom row, left to right: Seal Island zoom, small boat leaving Cape Town harbor, ship off Cape Town.

KHAYELITSHA (BEFORE AND AFTER APARTHEID)

New York, 2001 — Once Ikonos had begun selling imagery commercially, I purchased a 1-meter-resolution image of Khayelitsha in Cape Town in order to explore the latent possibility of a comparison between the Corona (before) and Ikonos (after) imagery, such that one could see the changes in one place across two moments in time. Most places in Cape Town had changed over the years — a lot had happened, needless to say, in South Africa between 1968 and 2000 — but none as significantly as the area around the highway that was photographed under construction by the Corona satellite in 1968.

In those thirty-two years, a city had emerged along that highway. Khayelitsha, which takes up about 43 square kilometers (16.6 square miles) of those sandy flats, was laid out in the early 1980s as a "new home" (its Xhosa name) for so-called "nonwhite" residents of the Cape Peninsula then living in what were defined as illegal shantytowns and squatter camps around the white city. Khayelitsha is a direct expression of the spatial politics of apartheid. It had grown considerably: by the time the Ikonos satellite passed overhead, it had around 400,000 inhabitants. That is roughly the number of people in Orleans Parish, but with an average of around ten thousand people per square kilometer (just over one-third of a square mile), mostly in single-story buildings, shacks, and shanties, it is significantly more dense.[31] Khayelitsha was first settled — "informally" in a sense, but one could also say by force — when residents of the Crossroads settlement, located alongside the Cape Town International Airport, were removed there in 1983 with a promise to legalize their residency status. More fled to Khayelitsha in 1986 when Crossroads erupted in violence. The city grew exponentially over the next few years. It was a planned town (or "township," in apartheid terminology), but its formality effectively meant formalized squatting in planned areas known as "sites and services" areas, rows of single, double, or quadruple plots with shared outdoor faucets and toilets.

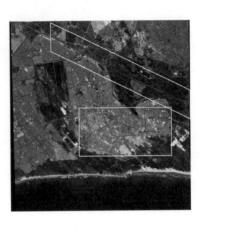

Zoom on area of interest: Khayelitsha and its environs, as seen by Corona KH-4B satellite in November 1968 (left), and by the Ikonos satellite in July 2000 (right).

Zoom on area of interest: N2 highway from
Cape Town to airport, as seen by Corona KH-4B
satellite in November 1968 (top), and by the
Ikonos satellite in July 2000 (bottom).

Zoom on area of interest: Khayelitsha, seventeen
years before it was named as such, as seen
by Corona KH-4B satellite in November 1968.
Scale: 1 pixel = approximately 2 meters.

Zoom on area of interest:Khayelitsha, as seen by the
Ikonos satellite in July 2000. Scale: 1 pixel = 1 meter.

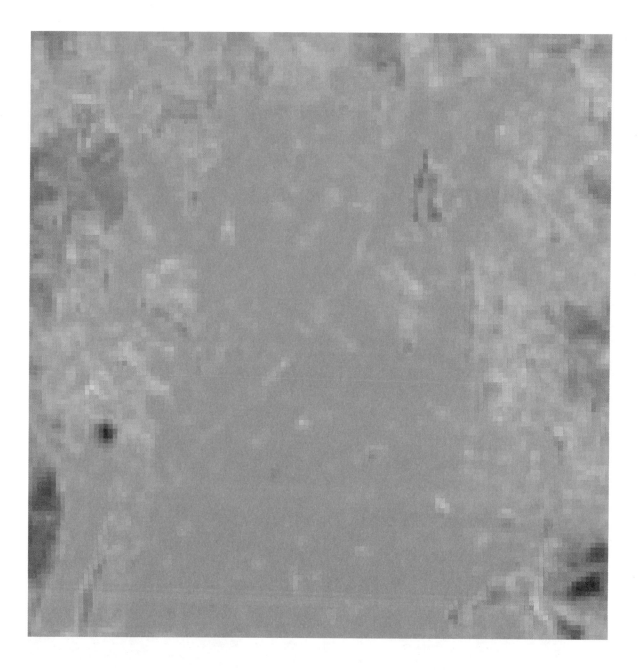

Zoom on area of interest: Pristina, Kosovo, as seen
by SPOT satellite on June 3, 1999. Scale of pixels
as displayed in gallery: 1 pixel = 20 meters.

4   Kosovo 1999: SPOT 083-264

*The necessity of linking satellite images
to the data that accompany their production*

New York, 2012—From the onset of their availability in the 1970s, the value of Landsat and SPOT image data was noticed more quickly and put to work more readily by environmentalists than by any other advocacy group. The view of Earth from outer space, from somewhere quite unprecedentedly not on the Earth, made the existence of the planet as such somehow more evident. Activists recognized the power of the image to explore and publicize oil spills and fires, toxic waste dumps, the effects of irrigation, acid rain damage to forests, tropical deforestation in the Amazon, shrinking lakes and seas, calving icebergs, or quite simply the vulnerability of the Earth as a small blue marble, floating in space.[32] People were absent from these images, but their effects could easily be seen—and seen as threatening— precisely because the effects were so large that they could even be captured at low resolution.

In 2000, with the launch of Ikonos, people still for the most part were absent, but finally the public could view the Earth with the same 1-meter resolution that the U.S. government had been using since the 1960s. The possible domains of "civic satellite surveillance" or "satellite imagery activism" opened up—activists were granted the clarity of vision to be able to identify burned villages in Darfur, nuclear sites in Iran, prison camps in North Korea, or the ruins of Grozny in Chechnya.

But the "humanitarian interventions" of the 1990s—that is, wars undertaken in the name of humanitarian values—had already brought satellite imagery into the international public imagination, and at the decade's end, with the war in Kosovo, the United States and NATO committed themselves decisively to large-scale releases of aerial and satellite images.

One set of pictures stood out for me in those days in the spring and summer of 1999: publicly released images of a newly dug mass grave in a place called Izbica. The first images were shown by NATO at a press conference in April, reporting a

1994
Time runs a digitally altered **LAPD mug shot** of O.J. Simpson, which makes him look sinister. The image raises ethical questions about how such technology should be used.

1996
New software allows thousands of photographs to be the building blocks of a single mosaic image. For its 60th-anniversary issue, LIFE runs a photomosaic portrait (of Marilyn Monroe) by Robert Silvers made up of **previous LIFE covers.** It is among the first magazines to feature such an image on its cover.

1999
Photographs taken by U.S. spy satellites of a **suspected mass grave site** last May (left), and again in June (right), indicate that the ground had been bulldozed—an apparent attempt by Serbian forces to erase evidence of a massacre in Kosovo. ●

U.S. spy satellite imagery featured in *Life* magazine's "Great Pictures of the Century and the Stories Behind Them."
LIFE MAGAZINE (OCTOBER 1999), P. 66

massacre and showing the grave. Then a second set was shown by the Pentagon in June, showing that the grave had been destroyed—no doubt in response to the first images. They were all, of course, part of a campaign to persuade the public that the air war over Yugoslavia was just and necessary. When, at the end of the year and the end of the century, *Life* magazine chose the "Great Pictures of the Century," the first Izbica image was featured as the single image from 1999 included in the "Flashbacks" section.[33] I thought the story and especially the images were worth pursuing.

Due to their classification (we still don't know which satellite took the picture), NATO and the Department of Defense released the images simply as a series of locked pixels with undetermined coordinates.[34] Where was the grave, precisely? Investigating the archive of SPOT image data collected throughout the eleven-week air war, I found only two cloudless days. This led me to guess that the military had taken the famous images on one of those days. In tandem with another image, taken by a German military drone and released by the Bundeswehr as part of its own public-relations campaign, I was able to deduce the longitude and latitude of the grave.

SPOT 083-264 insists on the necessity of linking satellite images to the data that accompany their production, not simply for technical reasons, but for ethical and political ones as well. Now, reinscribed with the data that created it, the image becomes a memorial to an evacuated violence.

KOSOVO: SPOT 083-264, JUNE 3 AND JUNE 6, 1999

Kosovo, 1999—Between March 22 and June 14, 1999, the commercial French SPOT satellites aimed their sensors at sites in southeastern Europe seventy-two times, collecting data on the ground from an altitude of 822 kilometers (511 miles). Thousands of megabytes of data about war, displacement, and destruction, because—not by accident—the satellites were passing over Kosovo. Their 10-meter and 20-meter resolution data were immediately stored and made available publically, directly from an active war zone, on almost every day of the NATO air campaign.

Here are data sets from June 3 and June 6, two of the rare cloudless days during the war, as the satellites recorded what was happening in the scenes below, gathering information on the landscape of ethnic cleansing and war. Permanent digital records, created at the speed of light: sixty square kilometers in a matter of seconds.

KOSOVO: SAT·K-J ID·DATE·TIME·CAMERA·SENSOR
SPOT scene: 2·083-264·99/06/03/·09:18:53·2·X
Cloud cover: 0%
Extents: 11,938,066 pixels
Top left: 43°01′12″ N, 20°39′55″ E
Bottom right: 42°23′40″ N, 21°23′02″ E
Coverage: 79.88 x 59.780 kilometers
Resolution: 1 pixel = 20 meters

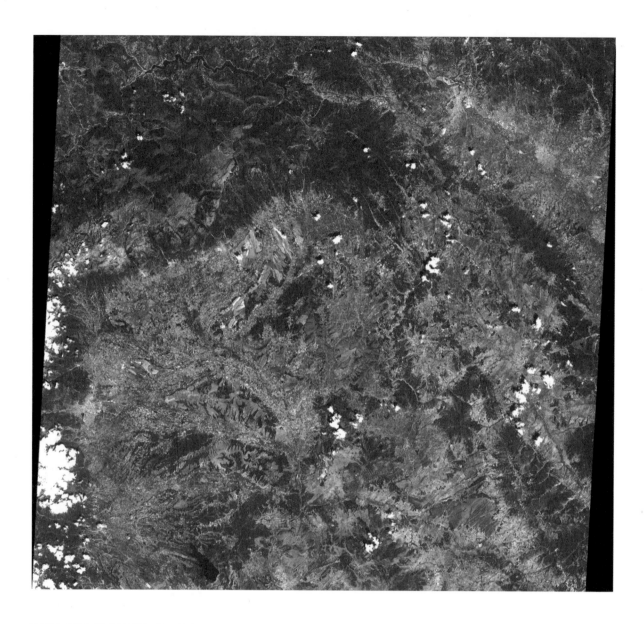

KOSOVO: SAT·K-J ID·DATE·TIME·CAMERA·SENSOR

SPOT Scene: 4·082-264·99/06/06/·09:32:30·1·M

Cloud cover: 0–10%

Extents: 38,834,152 pixels

Top left: 43°01′13″ N, 20°21′17″ E

Bottom right: 42°23′34″ N, 20°55′22″ E

Coverage: 64.81 x 59.92 kilometers

Resolution: 1 pixel = 10 meters

These are two scenes from the vast quantity of images which SPOT, Landsat, Sovinformsputnik, and other satellites record daily and store in databases: ready to be browsed and bought. They are collections of data—although they are presented as pixels that resemble an image—information waiting to be examined and interpreted, snapshots in time and space, this time of a war.

Other data were being collected at the time, as well, but we did not have quite the same access to it. "The former Yugoslavia is the most listened to, photographed, monitored, overheard and intercepted entity in the history of mankind," a U.S. State Department official had told the *New York Review of Books* a few years earlier in a story on the failure of U.S. intelligence agencies to make public the evidence of war crimes that it had collected through "what are officially called 'national technical means,'" and then what happened when they did. "A pair of photographs changed the course of the war in Bosnia," wrote Charles Lane and Thom Shanker about U-2 and satellite imagery from Srebrenica in 1995.[35]

By 1999, some products of that information-gathering enterprise did make their way into the public domain. This time, the U.S. and its NATO allies were in the fight, not trying to stay out of it. So Kosovo names, among other things, the conflict in which classified NATO images were finally released systematically to the public. And they were not simply pictures of the conduct of the war, but of its ostensible reasons. This time, in addition to footage of bombs and missiles, the public could see ethnic cleansing in progress: high-resolution imagery of mass graves, refugees in the mountains, burning villages, and organized deportations. It was the war in which satellite images were used as a way of forming public opinion. The manner in which they were released, however—as pictures, rather than as data—shows less the facts on the ground than the ability of the technology to record, in minute detail, these facts. No data, strictly speaking, were forthcoming at press briefings, and certainly not the data embedded in the pixels they had interpreted, and no information was available about the technology that had produced them. But there were lots of images. The Pentagon's military briefer, Major General Charles Wald, said one afternoon while answering a question about displaced people ("Do you have some recon on that?"): "So to answer your question, are we imaging them? Yes, we are." Then he added: "I won't talk about what kind of imagery that is."[36]

Stretched out horizontally across sixty kilometers, the Drenica Valley of Kosovo in early June is displayed in what the analysts of overhead imagery call "standard false color." We know how to read this image, more or less, because we know what the colors of the pixels conventionally represent: red is vegetation, purple is marshland or farmland, blue is roads, buildings, and bare soil, dark blue is clear water, white is clouds or smoke, and black is something burned. But add to this what else we know about these picture elements: that they present data. Each pixel designates 20 square meters—just over 215 square feet. Each one has an address, expressed in longitude and latitude, corresponding to a unique territory on the face of the Earth. And each one has a signature, the heat value of that place at the time the satellite passed silently above. That value is expressed as a number, which in turn has an assigned standard false color. The satellite gathers data— we see an image.

What can we see in this image data? It is the record of a war, not just of NATO's air war, but of the emptied cities and the burning villages, the refugees and what

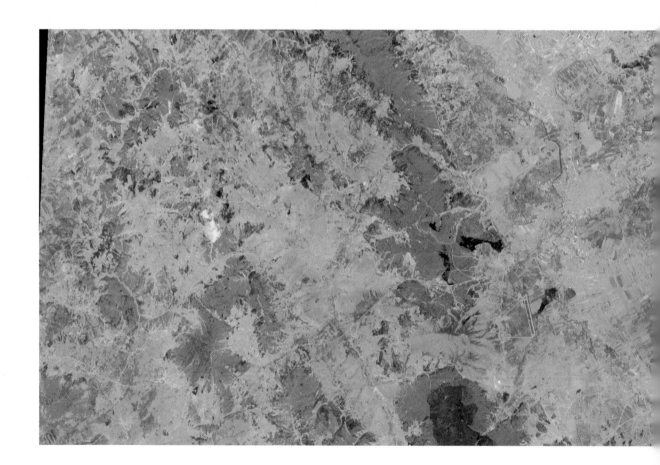

they've left behind, the mass graves and the crimes that are now too easily named with the term "ethnic cleansing." But what can we see? We know that a war is taking place and that it is presented to us in these colors that tell us something about a landscape. Red: at 20-meter resolution, the pixels hide the people who are hiding in the hills and forests. Blue: at 20-meter resolution, the small buildings in the villages are indistinguishable from the roads, and what remains is the large blue trace of the city, Pristina. To its west extends the Drenica Valley, where "the war started" and where some Serb fighters promised to end it.[37]

KOSOVO: SAT·K-J ID·DATE·TIME·CAMERA·SENSOR
SPOT scene: 2·083-264·99/06/03/·09:18:53·2·X
Zoom on area of interest: Drenica Valley, Kosovo,
June 3, 1999.

A DIGITAL MEMORIAL

The record of a double erasure, the evidence of a massacre and of a grave upturned, is digitized and remembered here, apparently by military surveillance satellites. The black is presented to us as the black of freshly upturned soil in the village of Izbica. Absorbing more heat than the adjacent grasslands, it is distinguished and recorded by the implacable sensors of the satellites. It would take only the next rain to wipe away the evidence, and then the grass would start growing again. But not on this image.

How can this image be located? In time and space, in history, in memory, or in a database? These—the remainders of the burial ground for scores of villagers in Izbica, killed by the military in March, buried by their neighbors soon after, and then removed from the scene in June—are just a few of the millions of pixels that make up this image, and there are certainly many more worth memorializing. But how can we return this picture to its rightful place in memory, realign it with the data stripped away from it as it became public? For now, we are left to our own devices.

It's not that hard to do. Izbica is a small village, locatable on a map with coordinates. In May, as imaged by a German military drone, the grave was just off to the side of a curved road. The road is blurred on the 10-meter SPOT data, but recognizable. And when rotated, the image of the tampered grave released by the Pentagon retrieves one element of the data, a north orientation. Compared with the Department of Defense image, the coarseness of the pixels on the SPOT data is deceiving— but the shape is the same, and nameable with longitude and latitude points on a map. It can be marked in digital space and time—as a memorial to an event which should not be erased.

"In recent weeks, refugees have reported that Serbian forces have undertaken mass executions and individual summary executions in at least 70 towns and villages throughout the province. Overhead imagery confirms the presence of mass grave sites in Pusto Selo and Izbica." FROM "ERASING HISTORY: ETHNIC CLEANSING IN KOSOVO," REPORT RELEASED BY THE U.S. DEPARTMENT OF STATE, WASHINGTON, D.C., MAY 1999, FIG. 2

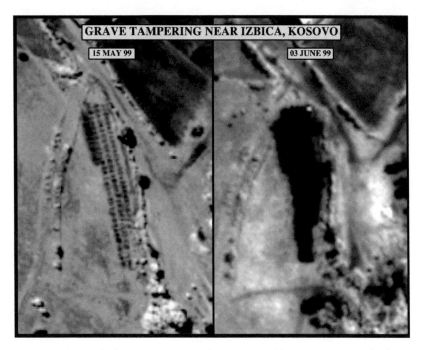

"Assessment photograph of grave tampering near Izbica, Kosovo, used by Assistant Secretary of Defense for Public Affairs Ken Bacon during a press briefing on NATO Operation Allied Force in the Pentagon on June 9, 1999." CAPTION AND PHOTO: U.S. DEPARTMENT OF DEFENSE

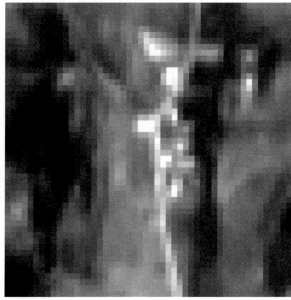

German CL 289 drone imagery of probable graves
at Izbica, Kosovo, May 1999. Originally published
on a Bundeswehr web page of "Drohnen-Bilder,"
reproduced at John Pike's Federation of American
Scientists archive, "Kosovo Operation Allied Force
Imagery," now at GlobalSecurity.org.

KOSOVO: SAT·K-J ID·DATE·TIME·CAMERA·SENSOR
SPOT Scene: 4·082-264·99/06/06/·09:32:30·1·M
Extents: 2,500 pixels
Top left: 42°43′47.81″ N, 20°38′11.62″ E
Bottom right: 42°42′9.47″ N, 20°41′7.87″ E
Coverage: 500 x 500 meters
Resolution: 1 pixel = 10 meters
Izbica, Kosovo, June 6, 1999

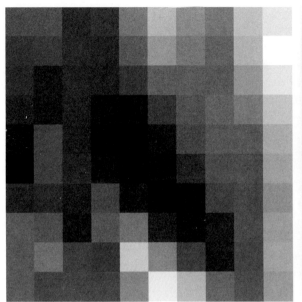

KOSOVO: SAT·K-J ID·DATE·TIME·CAMERA·SENSOR
SPOT Scene: 4·082-264·99/06/06/·09:32:30·1·M
Extents: 100 pixels
Top left: 42°43′45.76″ N, 20°39′37.96″ E
Bottom ight: 42°43′41.98″ N, 20°39′41.48″ E
Coverage: 100 x 100 meters
Resolution: 1 pixel = 10 meters
Grave site, Izbica, Kosovo, June 6, 1999

KOSOVO: SAT·K-J ID·DATE·TIME·CAMERA·SENSOR
SPOT Scene: 4·082-264·99/06/06/·09:32:30·1·M
Extents: 10,562,500 pixels, enhanced from 100 pixels
Top left: 42°43′45.76″ N, 20°39′37.96″ E
Bottom right: 42°43′41.98″N, 20°39′41.48″ E
Coverage: 100 x 100 meters
Resolution: enhanced to 1 pixel = 3 centimeters
Grave site, Izbica, Kosovo, June 6, 1999

CLOUDED MEMORY

New York, 2000—SPOT 083-264 was just one scene from the vast quantity of images that the commercial satellites of 1999 recorded daily and stored in databases. Here is a graphical snapshot of a tiny part of the SPOT database, counting the passage of the satellites over one particular piece of ground which during the NATO air campaign came into broader focus: a place to watch over for some, a target for others. For a little more than eleven weeks, it was examined with almost every technology available to the military, governments, civil society, and the news media. But this is a picture of what the satellites saw over ten years, the ten bloody years of the disintegration of Yugoslavia, which started in Kosovo long before much of the world even knew what is was or where it was located. The density of the graph intensifies in 1995 as the war in Bosnia ended and NATO peacekeepers arrived, and again at the end, when the war finally came to Kosovo.

Go to the SPOT archive and search the database.[38] You can search by longitude and latitude or by the path and row of the satellite orbit, if you know the pattern. You can refine your search by date and time and, most importantly, you can *limit* your search by the same categories and by the quality of the image. They will not sell you an image that does not have a rating of "E" for excellent image quality, and they recommend that you check the cloud coverage—optimally, less than 10 percent is the most desirable for seeing the data in the image.

The database, however, stores everything. And you don't have to limit your search. The archiving software inserts every image and then classifies it: even the glitches and the days when the sensors recorded mostly clouds. And so browsing the database, even at the low resolution of the thumbnails in the online database, yields a lot of clouds.

And since imaging satellites still cannot see through clouds, the SPOT archive also constitutes an accurate index of what the very-high-resolution military satellites could see, too. Clouds.[39] There were only three cloudless days during the NATO campaign.[40]

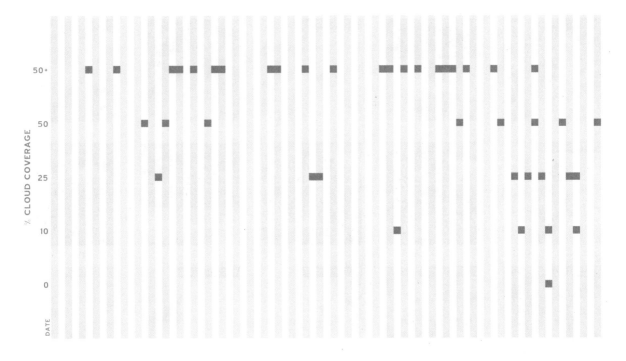

Availability of images, and percentage of cloud cover in them, for SPOT imagery of Kosovo, scene 083-264, March 24 – June 10, 1999. Each green square represents one image.

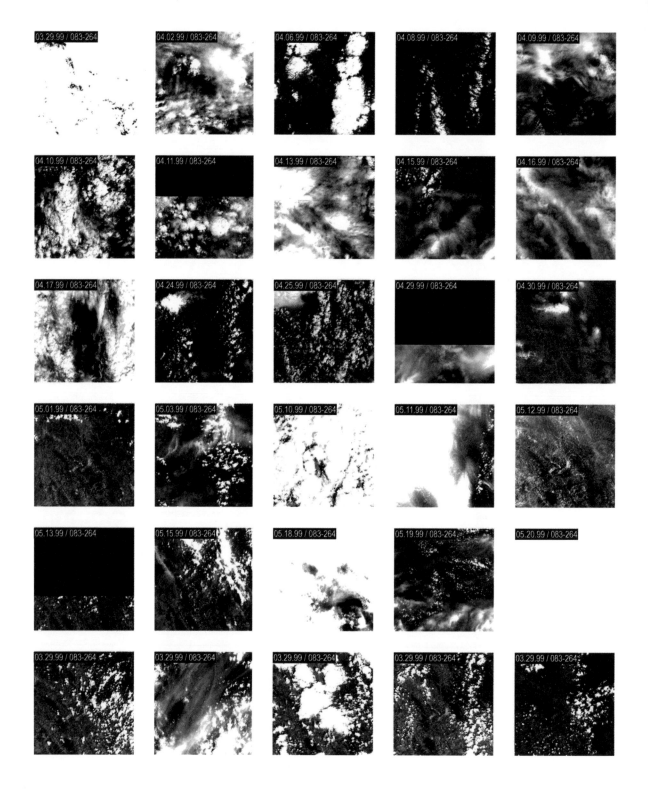

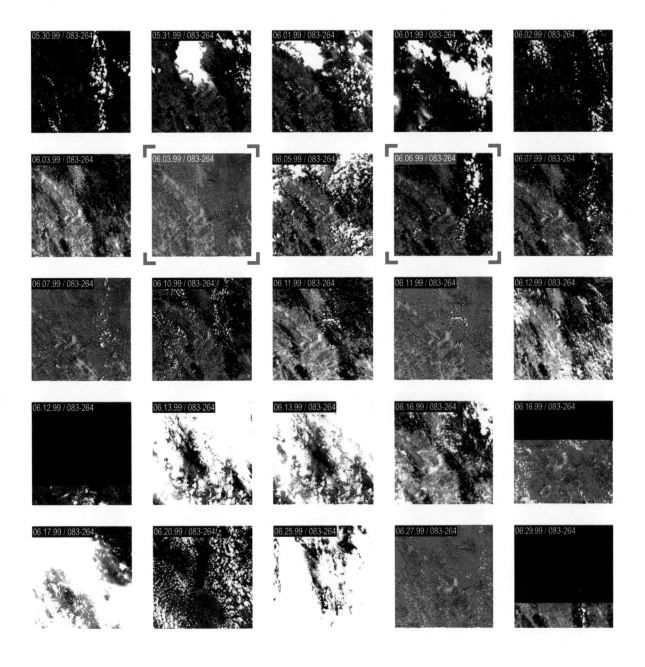

Fifty-five Quicklook thumbnails of SPOT imagery of
Kosovo, scene 083-264, March 29 – June 29, 1999.
Red brackets mark the images used in this project.

"This one-meter resolution satellite image of Manhattan, New York was collected at 11:43 a.m. EDT on September 12, 2001, by Space Imaging's Ikonos satellite. The image shows an area of white dust and smoke at the location where the 1,350-foot towers of the World Trade Center once stood. Ikonos travels 423 miles above the Earth's surface at a speed of 17,500 miles per hour." IMAGE AND CAPTION: GEOEYE, RELEASED SEPTEMBER 13, 2001

Lower Manhattan, September 15, 2001, as seen by Ikonos satellite. IMAGE: GEOEYE

5 New York, September 11, 2001

In a sense, I went from one mass grave to another,
but not intentionally

New York, 2012—I had been commissioned to produce an image for an exhibition on surveillance called *CTRL [SPACE]* at the ZKM, the Zentrum für Kunst und Medientechnologie, in Karlsruhe, Germany, in 2001. Ikonos had been launched the year before, and I had planned to order a 1-meter-resolution image of the Cameroonian rain forest. I looked almost every day for much of the year, but as the deadline approached, there were no cloudless days over that part of Cameroon, making it impossible to fill the order.

Then, the World Trade Center was attacked. A 1-meter-resolution Ikonos image of the site appeared in newspapers shortly thereafter, and I knew then what I would display at ZKM. I wanted to show Ground Zero a few days later, in its urban context—the empty streets that so many people came to remember from that time period—so that, like *SPOT 083-264*, the image could serve as a marker of an event.

The ZKM show opened in October 2001. The image was blown up and displayed on the floor of the gallery for people to examine, seventeen meters long and six meters wide. No one wanted to walk on Ground Zero. It was very raw.

The satellite image, available almost immediately, seemed to signify the realization of the promise of immediate public access to global satellite images, which is to say, the notion of "global transparency," that had been repeated with each new launch, but something different in fact happened. Access was denied.

The United States went to war very quickly in Afghanistan, and as CNN's new anchorman Aaron Brown put it on October 8, 2001: "It is not obviously a television war."[41] It was also not a transparent war. What was significant was the lack of imagery then—and the resolve of the Pentagon to enforce the lack. The Department of Defense purchased the exclusive rights to all Ikonos imagery over Afghanistan and its surroundings at the beginning of the war—for a month—and then renewed that contract for a second month. Even after that, though, things did not entirely clear up. Some time after this project, I attempted to purchase a QuickBird image from the day that the National Museum of Iraq was looted, but DigitalGlobe refused to sell it to me on the grounds that its release might endanger U.S. troops.

The attack on the World Trade Center was a very important event for architecture, but not simply because thousands of people died in the destruction of two tall buildings, and a number of others, as well. These were not simply buildings, and they were not attacked simply as buildings: they were images.

The attack was an event that also has left us with a lot of images—it was an event composed of a multitude of images, as Gilles Peress and his colleagues made so clear in their *Here Is New York* exhibition and archive.[42] My interest was in looking very carefully, as closely as possible, at just one of them. It has an architecture of its own, and it teaches us something about the architecture of the event—and also about the asymmetry of the state of images between New York and Afghanistan at that time.

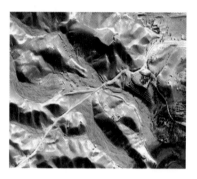

In October 2001, after U.S. and coalition airstrikes against al-Qaeda training camps north of Jalabad, Afghanistan, John Pike released December 1999 Ikonos imagery of the Darunta camp complex, including Osama bin Laden's Tora Bora base. Pike used earlier images, acquired shortly after the launch of Ikonos, for before-and-after comparisons with imagery released by the U.S. Department of Defense of the targeted sites.
IMAGE: IKONOS SATELLITE, COURTESY OF GEOEYE

Architects wanted to "respond" to the events of September 11 by building new memorials. For me, nothing was more superfluous. What we needed was to come to terms with the site itself, not to hide from it by building something else. When a reporter asked me about this a few months later, all I could do was point at the missing: "Memorials seem to be our instant answer to disaster, and that's why no one ever calls it a mass grave.... And has anyone really asked what it means to build a memorial when you are still in the middle of the war? I think the site itself is the memorial. This is a mass grave—the site is what it is."[43]

Having looked at a lot of imagery from Bosnia and Kosovo and then the scarred landscapes of apartheid South Africa, I guess I approached these images with a certain context in mind. In a sense, I went from one mass grave to another, but not intentionally. Graves simply need markers, and more than anything else, that's what the image did for me: it did not reveal a lot, it did not fill us with awe, it just marked the spot, one of the spots, where something happened.

NEW YORK, SEPTEMBER 11, 2001, FOUR DAYS LATER

New York, 2001—Saturday, September 15, 2001, 11:54 a.m. Between a satellite and thousands of bodies, a cloud of smoke drifts.

Space Imaging's Ikonos satellite takes a high-resolution snapshot from outer space of a city in a state of emergency. The satellite monitors the Earth's surface, collecting data. That Saturday morning, the cloud of smoke slowly drifts away from the disaster.

There is a lot to see in this picture, too much in fact. The density of its detail demands that it be viewed close-up. But there is no single thing to look for and no particular piece of evidence that tells the whole story. And so the entire image is on view here, blown up, too large to see all at once. But the zoom offers no revelation, no instant of enlightenment, and no sublime incomprehension, either. It tells many stories. What has happened? The satellite's sensors capture a mass grave, a record of a crime or an act of war. Unfortunately or fortunately, the image itself offers no instructions about how to understand or respond to what it has recorded in memory.

For the record, this image should not exist, and neither should the events it has captured. It is an unacceptable image, but it is imperative that we look at it.

Here are two 1-meter-resolution satellite images of the aftermath of the event: detailed pictures of a disaster. The first image was collected at 11:43 a.m. on September 12, 2001, a bit more than twenty-four hours after the attack, released by Space Imaging and published worldwide almost immediately.[44] The second, from September 15, was purchased at a cost of several thousand dollars and arrived in the form of a 323-megabyte data file from Space Imaging. It forms the basis of the images in this chapter.

High-resolution satellite images are one of our most powerful metaphors for the new condition of universality: an all-seeing image, potentially of any point on Earth, available to almost anyone, rich in data that can be used for purposes we cannot even predict. It offers precision, time-stamped evidence from an authoritative eyewitness. But it wants to represent the magnitude of the event: with the sublimity appropriate to a catastrophe, it offers the view from above, from "overhead," in which the city is seen in the midst of an emergency. It tries to see everything at once, everything that cannot usually be seen with the human eye. If it fails, it should tell us—in just the manner proper to the sublime—about the limits of our understanding. In the end, though, the image is neither the definitive eyewitness nor the record of our incomprehension.

The buildings are missing, disintegrated into a vast zone of ruin. The city is quiet, except for intensive activity around the site. There are trucks along

Zoom on area of interest: emergency vehicles on the West Side Highway at West 12th Street, as seen by Ikonos satellite, September 15, 2001. Actual scale of pixels as displayed in the gallery: 1 pixel = 1 meter.

ominously empty highways, removing the debris. New York City's rigorous urban grids are broken up by the shadows of the buildings that remain, but also by the dust and smoke and the rubble of the very large buildings that have collapsed. At 11:54 in the morning, four days later, says the image, this is what it all looked like.

But the image offers only a certain kind of evidence. When the pixels finally reveal themselves as simply the pixels that make up the image, they are as silent as what they are picturing. This evidence tells little and is of little use, forensically. In their matter-of-factness, the pixels will stay, here on this image, even as the debris is removed, day by day, from the site. At least we will always be able to locate the rubble here.

So if in fact transparency is trivial, and nothing new is discovered about the event, we must rather say: here it is, the event is encoded right here, by the light that has travelled from the ground to the satellite, captured in an instant as the memory of this event. As data. Mutable, yes, but no less a memory, all the same.

What is missing from this image is what is missing from the city or the world, and it is always missing at the limits of 1-meter resolution, for all its detail. What is

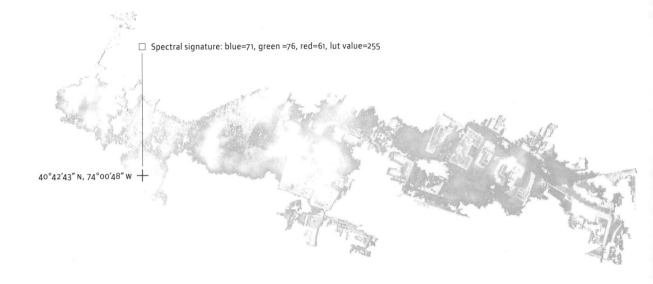

Spectral signature: blue=71, green =76, red=61, lut value=255

40°42′43″ N, 74°00′48″ W

missing are the missing, thousands of people missing, nearly three thousand ultimately confirmed dead. Beneath or beyond the limits of visibility, of data, are the dead. And yet they remain in the image, in the ruin of the image, and demand a certain care or respect.

In the image, four days later, the ruin is still on fire. Smoke hovers nearby, displaced from the site by the wind. It does not cast a shadow, the way a cloud farther to the south obscures the area near the rubble of the World Trade Centers. During the weeks following September 11, one could not always register directly what had happened in the city—until the wind changed directions and you smelled the smoke.

It is hard to isolate anything on this image. When one tries to isolate the disaster site by selecting similar pixels, the image-processing software tends to equate rubble with buildings. But it can isolate the smoke and what remains hazily below the smoke. So choose a pixel in the middle of the disaster site—it has a longitude and a latitude and a spectral signature. The software can then associate this pixel with similar pixels, and the area can grow to define the most changeable part of the site: the cloud of smoke that bears witness to the crime. Displaced, caught in motion, it records a particular moment of September 11, four days later.

WTC site | Time: 15:54 GMT | Acquisition Date: 2001-09-15 | IKONOS Satellite | Southeast Long/Lat: 73°59'25.11"W/40°40'15.52"N | Southwest Long/Lat: 74°01'08.26"W/40°40'15.52"N

"New York, September 11, 2001, Four Days Later,"
digital print from Ikonos satellite data of September
15, 2001 by Space Imaging. Scale: 1 pixel = 1 meter.
Full printed image measured 17 by 6 meters of
floor space.

Zoom on area of interest: World Trade Center site,
as seen by Ikonos satellite, September 15, 2001.
Scale: 1 pixel = 1 meter.

Zoom on area of interest: downtown Manhattan,
parts of Brooklyn and Queens, the eastern edge
of New Jersey, and the New York Harbor, including
Governors Island and the Statue of Liberty,
as seen by Ikonos satellite, September 15, 2001.
Scale: 1 pixel = 1 meter.

The changing perimeter of Ground Zero
from September to December 2001. Top
row: September 11, 14, and 19; bottom row:
September 27, October 24, and December 5.

6 Around Ground Zero

We needed not only to make a claim for a right to look,
but also to help realize it

New York, 2012 — In the year following the attacks on New York and Washington, there was a lot of critical commentary in the American press about the lurid voyeurism of gawkers at Ground Zero. The message seemed to be that people should not visit the site. City authorities wanted to erect a forty-foot opaque fence around it to keep the unauthorized out. I was part of a group of architects called New York New Visions that fought for a transparent fence, a permeable border through which people could view the scene (we called it "bearing witness") and pay respect to the dead. In a city where nothing of this scale had happened before, it seemed important that people did look and that looking become a sanctioned activity. I thought that we needed not only to make a claim for a right to look, but also to help realize it.

One of our strategies was to make it easier for people to get to and around the site. *Around Ground Zero* was a series of fold-out maps of the Ground Zero site. Working with others at New York New Visions and a team of volunteers, we printed three versions, updating them as the site changed, first in December 2001, then in March and again in June 2002. We distributed them on the streets around Ground Zero for free.

I was inspired in this by the work of a group in Bosnia called FAMA. In the 1990s, during the war and the siege of their city, they created hand-drawn and mass-reproduced tourist maps of Sarajevo. They charted buildings that were destroyed, sniper locations, and the sites of significant events in the war — partly as an ironic critique of the way the war in Sarajevo had become a matter of spectacle and fly-in, fly-out tourism, but also seriously, as a record of and a guide to a place that could often seem rather confusing or disorienting. I had visited Sarajevo in 1999, after the war, and used the map to navigate that city.

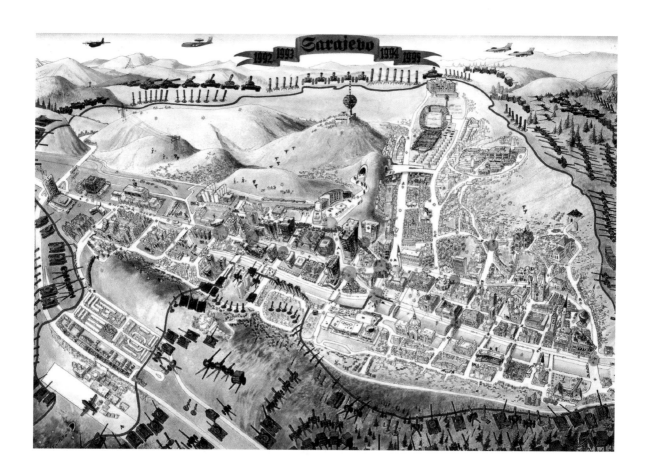

FAMA, "Survival Map, 1992–1996," Sarajevo, 1996.

THE MAP

New York, 2001–2002 — GROUND ZERO (also referred to as the inner zone, the pile, the pit, the site, the zone) is where the twin towers and the neighboring buildings of the World Trade Center collapsed, leaving a 16-story pile of debris and fragments of the exterior structural walls. The pile, now removed, consisted primarily of dust made up of the pulverized contents of the towers, extending deep into the underground base of the complex. As this foundation is excavated, "the bathtub," a 70 foot-deep slurry wall holding back the Hudson River and dating from the construction of the towers, has been exposed at the western side of the site.

The site is accessible to construction and rescue workers, firefighters and police involved in these efforts, public officials and the press, and others with special authorization. Recovery and removal operations go on 24 hours a day.

As of early March 2002, some 3,000 people are thought to have died at Ground Zero, including the 147 passengers aboard the two planes that struck the towers. The debris is removed carefully to allow for the identification of the missing. Ambulances bearing the remains of the victims are increasingly rare; a police escort may indicate that they are carrying the body of a firefighter, police officer, or emergency worker.

Very large CRANES, typically used in mining operations, were used to remove most of the debris, and in the process became a new landmark for downtown Manhattan.

TRUCKS carrying debris from the site leave it frequently. The rubble is taken to Piers 6 and 25 and loaded onto BARGES for shipment across New York Harbor to the Fresh Kills landfill on Staten Island, where it is reexamined by investigators and engineers. What remains of the steel structure is transported to Brooklyn for examination and then to scrapyards in New Jersey for recycling. The debris is estimated to weigh 1.35 million tons, about two thirds of which had been removed by March.

There is no PUBLIC ACCESS to Ground Zero beyond the "red line." Although it runs through buildings and other obstacles, it is visible as a chain-link fence covered in green fabric. Several construction access gates at major streets are guarded by police or National Guard troops. Between the fence and inner zone are areas used for staging, storage, construction and emergency vehicles. The GREEN FENCE prevents visitors from interfering with recovery and demolition work and ensures the safety of the public. The temporary barriers can be moved to allow for different conditions on a daily basis. Though segments of the fence obstruct the view, visitors have appropriated them as sites of memory and witness.

The openings afforded by cross streets and avenues often allow **VIEWS** of the site. As the demolition progresses, these locations change. There are two official **VIEWING PLATFORMS**. Just inside the site at the southwest corner, the Port Authority platform has been used by dignitaries, celebrities, officials, and by the victims' families, many of whom have written messages on the handrails. A public viewing platform has been built on Fulton Street, near Broadway. It too has become a repository of messages left by visitors. Tickets to the viewing platform are distributed daily at the South Street Seaport Museum's ticket booth at Fulton and South Street on Pier 16.

In addition to the collapsed WTC buildings there were some 45 seriously damaged buildings around the site. Many of these have been repaired and reopened. Some of these structures have been temporarily covered by construction **CURTAINS** to prevent injury and damage from falling debris.

Within and around the site, many spontaneous **MEMORIALS** to the dead and missing have been constructed, disassembled, lost, removed, moved, or rebuilt since September 11. At Ground Zero, large pieces of the building, notably the remnants of the walls and fragments of its steel structure, served as markers of the catastrophe until mid-December. At the fence, the viewings platforms and elsewhere, memorials have included candles, photographs, flowers, flags, messages, and teddy bears, which have become the most prominent symbols of mourning and memory. Two memorials were created along the Battery Park Esplanade, one for uniformed officers and the other (a wall of hundreds of teddy bears, now largely removed) for civilian victims. The fence along St. Paul's Chapel on Broadway at Fulton Street has been a constant, if changing, memorial site. An unclaimed bicycle on Broadway at Cedar Street serves as a memorial to the unknown number of undocumented immigrant workers who died on September 11. Thousands of flyers with the photographs of the missing, their names and descriptions, were posted at hospitals, rescue centers, bus stops, and phone booths around the City. They first appeared as signs of hope, and later became markers of loss and memory.

The "Tribute in Light," a projection of two beams of light into the night sky over Battery Park City, is scheduled to run from March 11 through April 13. Another temporary memorial, composed of the remnants of a destroyed sculpture from the WTC plaza, is planned for Battery Park.

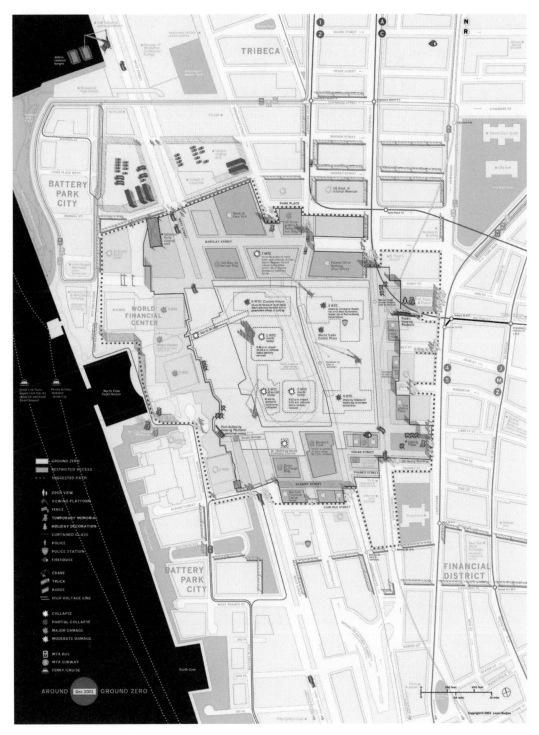

"Around Ground Zero," December 2001 (front)

Wall of the missing, St. Vincent's Hospital, September 2001. Photograph by Margaret Morton.

Around Ground Zero

Developed as a project of the Temporary Memorials Committee of the New York - New Visions Coalition for Rebuilding New York.

Concept design: Laura Kurgan.
Project team: Laura Kurgan, Janette Kim, Bethia Liu, with Rivka Mazar and Donald Shillingburg.
Photography: Margaret Morton.

Copyright 2001 Laura Kurgan.

Ground Zero (also referred to as the inner zone, the pile, the pit, the site, the zone) is where the twin towers and the neighboring buildings of the World Trade Center collapsed, leaving a 16-story pile of debris and fragments of the exterior structural walls. The pile consists primarily of dust made up of the pulverized contents of the towers, extending deep into the underground base of the complex. The site is accessible to construction and rescue workers, firefighters, and police involved in these efforts, as well as to public officials and the press, and others with special authorization. Coordinated by four construction companies, workers are engaged in removal and recovery operations 24 hours a day.

As of mid-December, some 3,000 people are thought to have died at Ground Zero, including the 147 passengers aboard the two planes that struck the towers. The debris is removed carefully to allow for the identification of the missing. Ambulances, bearing the remains of the victims still being unearthed, depart the site with some regularity; a police escort may indicate that they are carrying the body of a firefighter, police officer, or emergency worker.

The **cranes** on the site have become a new landmark for downtown Manhattan. Typically used in mining operations, they are much larger than ordinary construction cranes. **Trucks** carrying debris from the site leave it frequently.

Debris hauled from the site is taken to Piers 6 and 25 and loaded onto **barges** for shipment across New York Harbor to the Fresh Kills landfill on Staten Island, where it is re-examined by investigators and engineers. What remains of the steel structure is transported to Brooklyn for examination and then to scrapyards in New Jersey for recycling. The debris is estimated to weigh 1.2 million tons, about half of which had been removed by mid-December.

There is no **public access** to Ground Zero beyond the 'red line.' Although it runs through buildings and other obstacles, it is visible as a chain-link fence covered in green fabric. Several construction access gates at major streets are guarded by police or National Guard troops. Between the fence and the inner zone are areas used for staging, construction, and emergency vehicles. This **green fence** prevents visitors from interfering with recovery and demolition work and ensures the safety of the public. The temporary barriers can be moved to allow for different conditions on a daily basis. Though segments of the fence obstruct the view, visitors have appropriated them as sites of memory and witness.

The openings afforded by cross streets and staging areas allow **views** of the site. As the demolition progresses, these locations change. There are two official **viewing platforms**. Just inside the site at the southwest corner, the Port Authority platform has been used by dignitaries, celebrities, officials, and by the victims' families, many of whom have written messages on the handrails. A public viewing platform has been built on Fulton Street.

Within and around the site, many spontaneous memorials to the dead and missing have been constructed, disassembled, lost, removed, moved, or rebuilt since September 11. At Ground Zero, large pieces of the building, notably the remnants of the walls and fragments of its steel structure, served as markers of the catastrophe. Most of these have disappeared. Around the site, at the fence, the viewing platforms and elsewhere, memorials have included candles, photographs, flowers, flags, messages, and teddy bears, which have become the most prominent symbols of mourning and memory. Two memorials were created along the Battery Park Esplanade, one for uniformed officers and the other (a wall of hundreds of teddy bears, now largely removed) for civilian victims.

Tens of thousands of flyers with the photographs of the missing victims of the disaster, their names and descriptions, were posted near hospitals, rescue centers, bus stops, and phone booths around New York City. These flyers first appeared as signs of hope, and later became markers of loss and memory.

Elsewhere, memorials have also been created at many firehouses and police stations, Washington Square Park, Union Square Park, the Lexington Ave. Armory, Bellevue and St. Vincent's Hospitals, Grand Central and Pennsylvania Stations, and the Family Assistance Center on Pier 94.

"It is a burial ground. It is a cemetery, where the men and women we loved are buried. Where they rest is now hallowed ground."

AROUND Dec 2001 GROUND ZERO

"Around Ground Zero," December 2001 (back)

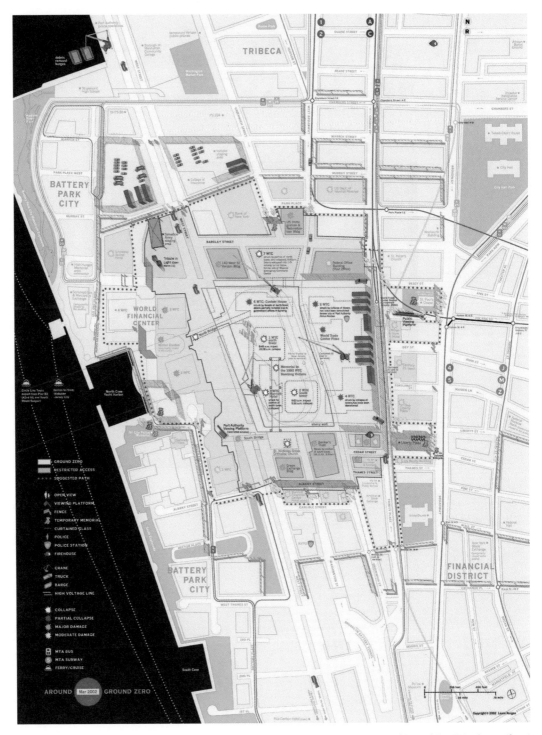

"Around Ground Zero," March 2002 (front)

Wall of the missing, St. Vincent's Hospital, September 2001. Photograph by Margaret Morton.

Around Ground Zero

Ground Zero (also referred to as the inner zone, the pile, the pit, the site, the zone) is where the twin towers and the neighboring buildings of the World Trade Center collapsed, leaving a 16-story pile of debris and fragments of the exterior structural walls. The pile now removed, consisted primarily of dust made up of the pulverized contents of the towers, extending deep into the underground base of the complex. As this foundation is excavated, "the bathtub," a 70 foot-deep slurry wall holding back the Hudson River and dating from the construction of the towers, has been exposed at the western side of the site.

The site is accessible to construction and rescue workers, firefighters and police involved in these efforts, public officials and the press, and others with special authorization. Recovery and removal operations go on 24 hours a day.

As of early March 2002, some 3,000 people are thought to have died at Ground Zero, including the 147 passengers aboard the two planes that struck the towers. The debris is removed carefully to allow for the identification of the missing. Ambulances, bearing the remains of the victims are increasingly rare; a police escort may indicate that they are carrying the body of a firefighter, police officer, or emergency worker.

Very large **cranes**, typically used in mining operations, were used to remove most of the debris, and in the process became a new landmark for downtown Manhattan.

Trucks carrying debris from the site leave it frequently. The rubble is taken to Piers 6 and 25 and loaded onto **barges** for shipment across New York Harbor to the Fresh Kills landfill on Staten Island, where it is reexamined by investigators and engineers. What remains of the steel structure is transported to Brooklyn for examination and then to scrapyards in New Jersey for recycling. The debris is estimated to weigh 1.35 million tons, about two thirds of which had been removed by March.

There is no **public access** to Ground Zero beyond the "red line." Although it runs through buildings and other obstacles, it is visible as a chain-link fence covered in green fabric. Several construction access gates at major streets are guarded by police or National Guard troops. Between the fence and the inner zone are areas used for staging, storage, construction and emergency vehicles. The **green fence** prevents visitors from interfering with recovery and demolition work and ensures the safety of the public. The temporary barriers can be moved to allow for different conditions on a daily basis. Though segments of the fence obstruct the view, visitors have appropriated them as sites of memory and witness.

The openings afforded by cross streets and avenues often allow **views** of the site. As the demolition progresses, these locations allow two official **viewing platforms**. Just inside the site at the southwest corner, the Port Authority platform has been used by dignitaries, celebrities, officials, and by the victims' families, many of whom have written messages on the handrails. A public viewing platform has been built on Fulton Street, near Broadway. It too has become a repository of messages left by visitors. Tickets to the viewing platform are distributed daily at the South Street Seaport Museum's ticket booth at Fulton and South Street on Pier 16.

In addition to the collapsed WTC buildings there were some 45 seriously damaged buildings around the site. Many of these have been repaired and reopened. Some of these structures have been temporarily covered by construction **curtains** to prevent injury and damage from falling debris.

Within and around the site, many spontaneous **memorials** to the dead and missing have been constructed, disassembled, lost, removed, moved, or rebuilt since September 11. At Ground Zero, large pieces of the buildings, notably the remnants of the walls and fragments of its steel structure, served as markers of the catastrophe until mid-December. At the fence, the viewing platforms and elsewhere, memorials have included candles, photographs, flowers, flags, messages, and teddy bears, which have become the most prominent symbols of mourning and memory. Two memorials were created along the Battery Park Esplanade, one for uniformed officers and the other (a wall of hundreds of teddy bears, now largely removed) for civilian victims. The fence along St. Paul's Chapel on Broadway at Fulton Street has been a constant, if changing, memorial site. An unclaimed bicycle on Broadway at Cedar Street serves as a memorial to the unknown number of undocumented immigrant workers who died on September 11. Thousands of flyers with the photographs of the missing, their names and descriptions, were posted at hospitals, rescue centers, bus stops, and phone booths around the City. They first appeared as signs of hope, and later became markers of loss and memory.

The "Tribute in Light," a projection of two beams of light into the night sky over Battery Park City, is scheduled to run from March 11 through April 13. Another temporary memorial, composed of the remnants of a destroyed sculpture from the WTC plaza, is planned for Battery Park.

A project of the Temporary Memorials Committee of New York New Visions Coalition for Rebuilding New York

Concept and Design: Laura Kurgan
Project team: Janette Kim and Belma Liu with Nuha Mazer and Donald Shillingburg. Photography: Margaret Morton. Printing: AGW Lithographers, Inc. This map is printed on Sappi PotSdrlatch 80 text.

Support from the following organizations and corporations is gratefully acknowledged: American Institute of Architects, New York Chapter; Con Edison; The New York Community Trust; The New York Foundation for the Arts; The Open Society Institute; Princeton University; TSRKK Charit\>Day; The Van Alen Institute; and two anonymous donors.

For information: **http://www.aroundgroundzero.net**
Web design: Razorfish
Web hosting: Logicworks

COURTESY FEMA/MATRE

"It is a burial ground. It is a cemetery, where the men and women we loved are buried. Where they rest is now hallowed ground."

AROUND GROUND ZERO

Mar 2002

INTERVIEW WITH ALICE TWEMLOW

New York, 2002 —

AT: What was the stimulus for your mapping project?

LK: The project grew out of discussions in New York New Visions, a temporary coalition of architects and designers who gathered to imagine responses on an urban scale to September 11. I was working in a group interested in proposing ideas for temporary memorials to the events. I had a memorial map of Sarajevo — designed by FAMA, the same activist designers who'd created the *Survival Guide Sarajevo* early in the siege — that I had used to tour the city after the war in 1999.[45] I brought it into one of our meetings to provide inspiration and a bit of context because I was trying to show that memorials don't have to be permanent.

AT: Why a map?

LK: It became clear that at that particular moment in time, the most pressing need was to respond to the number of people who were going to Ground Zero and looking at what was there. When I was first at the site, visitors had no idea which streets were open or closed and no idea where the towers had been. They crowded around these little Xeroxed FEMA maps that were posted on walls for the construction workers. Often, the gate guards had enlarged the FEMA maps and had them leaning against the back of a chair to avoid answering people's questions. There seemed to be an urgent need for a type of map that would help people make sense of what they were seeing, to orient themselves, in all senses of the word, or, if that was asking too much, at least one that would let them measure their disorientation in the face of the unimaginable. The site around what was the World Trade Center was manifestly disorienting for obvious reasons, and in a sense, that was as it should be. The map tried to address the unnecessary confusion and allow visitors to begin to take stock of what had happened.

AT: How did you make the map?

LK: I thought that we would simply be able to go to various agencies, such as the Department of Design and Construction, and ask them for relevant maps that they were drawing and using to put up fences and so on. But we found that just about every map being drawn in the city was either classified (still being used in the investigation or recovery effort) or otherwise not available for public use, and so we were left to our own devices. This meant we had to walk around and document where every fence, sign, Con Ed electric cable was and then draw it on the map. A lot of people helped out (especially Janette Kim and Bethia Liu from Princeton,

New and transparent fence around World Trade Center Site, with explanatory text of where the former buildings were located, among other details.   PHOTO: BETHIA LIU

the photographer Margaret Morton, and Rivka Mazar). We made lists, e-mailed notes, took photographs, followed the news in the papers and on television, talked to the cops and the workers, noticed as subways and streets closed and opened, collected everything and anything we could. The data gathering itself was a very low-tech process. The map is a compilation of all this ground-level surveying and then a number of digital sources.

There were many debates in the group about what should be emphasized and what should be on or off the map. As with any map, this one has a point of view. We tried to be factual, not emotional, open-ended, rather than didactic, without hiding the political and ethical stakes of the project. There was a long time when we had images of the burning towers on the front fold of the mockups. We took them off in the end to avoid spectacularization. We also avoided the images that had already become too familiar.

The map, though, was not just a humanitarian effort. We wanted to insist that the task of marking and reimagining the site had to be an open, public process, that the ordinary people who were visiting the site in such numbers should have access to the area and to information about it. The map says that, and it tries to enable it, as well. We wanted to underline the fact that before the "real" debate about memorials and buildings had even begun, people were already making their own tributes, markers, and memorials. We made a point of noting that the site was a cemetery, a mass grave—an open tomb of unknowns. And we wanted to begin recording the fact that a new history began on September 11 and that one of its sites was around Ground Zero.

AT: In what ways did Ground Zero transform from a disaster site to a tourist destination?

LK: When I began working on the map, the loudest voices were opposed to tourism in relation to the disaster—after all, they said, the site was still an open grave.[46] I wanted to say, "You should allow tourists down there because they are not doing anything bad—just bearing witness or paying respects or looking at the site is an important part of coming to terms with what has happened." It was a kind of an activist project on my part.

Then the situation changed a lot. After six months, "the bathtub" (the seventy-foot-deep slurry wall that holds back the Hudson River) was exposed, and all of the debris and remains had been carted away. By May 29, 2002, the site had been officially converted from a site of recovery to a site of construction. By the time of the first anniversary, tourism had been completely embraced by the Port Authority.

AT: And how did design play a role in this metamorphosis?

LK: As tourism began to be embraced, the perimeter fence became more and more of a focus, with graphic signs pointing out the "Best Views." In fact, something as simple as signage and the way-finding system did make a considerable difference in people's relationship to the site and the events. So design enabled this transformation, and it also became one of the biggest issues at stake in the public debates. Architectural renderings of what would become of the recovery site were on the front page of the *New York Times*, every night on CNN, and the topic of a huge town meeting and an ongoing public discussion, however imperfect.

Even though it didn't become an official map, perhaps it played some part in converting those who were initially suspicious about tourism. I hope it helped them to see another frame for looking at the site.

Now I think we've reached the point where we should be very careful about how we use signs and symbols in and around Ground Zero. I supported the fence becoming transparent (there were plans to build a forty-foot-high wooden fence around the site), but I don't like the interpretation of the event it seems to offer for tourists. I think that Ground Zero is something that demands great care and rigor in the way we frame it.

For instance, today, the fence announces a "Wall of Heroes," but who is defining whom as heroes? The apparently automatic patriotism of that wall of names seems inappropriate. Most of the names there are not heroes in the strict sense, but simply people who died doing a very ordinary thing: going to work one day. It does a disservice to the heroes to declare that everyone's a hero. The new fence is somehow at once watered-down and inoffensive and terribly didactic—not only telling you what is there, but also what to think about it. If we are framing a view of that site, we should be very careful with every word and image that is put there. In my opinion, there should be no words on that fence.

AT: I think that one of the ways in which your work differs from memorializing efforts, such as the "Wall of Heroes," for example, is due to your interest in data-driven work that is not melodramatic and not nostalgic. How does this mapping project relate to your other work and to other maps?

LK: I do a lot of work with mapping, but it is usually at the high-tech end of the spectrum, using Global Positioning System devices and high-resolution satellite imagery. Most of the other maps I have done have been consciously about disorientation: about how impossible it is to orient yourself in the new spaces of electronic technologies and also how important it is to use these new technologies for good ends, rather than the militaristic ones for which they were invented.

Although the Ground Zero project is not directly linked to my other work about maps and digital technologies of mapping—it is the most low-tech project I have done and the least disorientating—there are thematic links. A lot of my work has been about the major political events of the last decade—particularly military ones—and about reclaiming images of war as images of memory. Specifically, I have constructed what I like to call "digital memorials" with images generated by satellites. These images are snapshots in space and time, and I have tried to watch places such as Kosovo and the war crimes tribunal in The Hague to understand what difference new technologies can make for memory.[47]

Around Ground Zero is different, too, from the FAMA Sarajevo map, the main function of which was a one-time memorial. This project is time based. It is supposed to be a document of what was there at a given time—the temporary structures, the graffiti, the spontaneous memorials, and the shifting access routes—the things that would be erased in short order. You're right about the resistance to melodrama and nostalgia: my projects do have a tendency to be about scenes of destruction and yet to insist that there's no reversing the process. No map is going to undo what happened to the villagers in Izbica (Kosovo). There is something sober about simply marking where they were killed and buried and where the people who killed them came back to get rid of the evidence. In a way, I aim to emulate the unflinching eye of the satellite sensors, which took note of the graves and then of the mark on the Earth where the graves had been. The important thing is to do this without pathos: to engage in the act of bearing witness, of remembering and of testifying, simply because something happened that should not have happened. I think something of this was at work in the *Around Ground Zero* map, as well.

AT: How has the map's function evolved over the past year?

LK: At first, the imperative for the map was about memorializing through holding on to the ephemera that were a part of the process of coming to terms with what was going on. Now a different kind of information strategy is necessary for the rebuilding phase. I think that people know very little about what is happening, how the decisions get made and who sets the terms for the competitions. There is very little debate about what an architectural, political, and economic response should be to this act of terrorism. That's what the map is about now.

AT: What are your thoughts about the act of memorializing, both in relation to Ground Zero and in a larger social context?

LK: It is not only the way that graphic design has been used around the fence to interpret the event that is problematic, but what is being built on the site, too. It seems to me that the only options being made available to architects are symbols of recovery or triumph for a city that has overcome terrorism. There's still a battle between memory and money that characterizes the debate downtown. I think it's really scary.

I've done a number of projects about memory, memorials, and disasters in other countries. I've started becoming a little critical of the whole memory machine. Memorials have, I think, become instant answers to disasters, whether they are battles, or dictatorships, or tragedies such as Oklahoma City and Columbine. I think we should stop and think a little longer before building anything. The paradigm for memorials stopped with the Maya Lin Vietnam Memorial, and now, with no new ideas but an ever-increasing desire for monuments, it just gets repeated and repeated in different, all too often inappropriate contexts.

So far, the architectural proposals for rebuilding downtown—both the buildings and the memorials—appear to be abiding by the rules. I hope the new ones break the rules. It seems to me to do this project correctly, someone needs to break the rules.

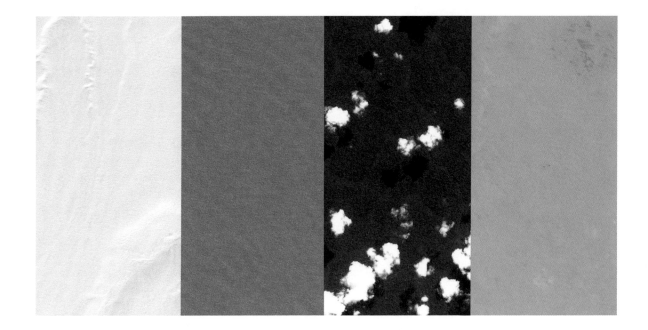

WHITE

1002 area, Arctic National
Wildlife Refuge (ANWR),
near Kaktovik, Alaska

Acquired: April 7, 2003,
21:12:49 GMT
Center coordinates:
69°47′ 59.46″ N, 144°32′ 33.9″ W
QuickBird satellite,
0.61 m per pixel

BLUE

Atlantic Ocean, intersection
of the Equator and the prime
meridian, south of Accra (Ghana)
and west of Libreville (Gabon)

Acquired: May 16, 2003,
10:19:54 GMT
Center coordinates:
0°0′7.02″ N, 0°0′1.62″ E
QuickBird satellite,
0.61 m per pixel

GREEN

Old-growth tropical lowland rain
forest, Cameroon, about 100 km
west of Yokadouma and 70 km
east of the Dja reserve

Acquired: December 4, 2001,
09:48 GMT
Upper left coordinates:
3°13′9.804″ N, 14°12′ 27.72″ E
Ikonos satellite,
1.0 m per pixel

YELLOW

Southern Desert, southeastern
Iraq, between Al Busayyah
and An Nasiriyah

Acquired: March 30, 2003,
07:32:10 GMT
Center coordinates:
30°18′48.96″ N, 46°22′25.68″ E
QuickBird satellite,
0.61 m per pixel

7 Monochrome Landscapes

My attention was then attracted by brighter colors
and by other sorts of contested territories

New York, 2012 — In the cloudy images from Kosovo, I had stumbled on the fact that something abstract was happening in the databases I examined. My attention was then attracted by brighter colors and by other sorts of contested territories. Included in a show called *Architecture by Numbers* at the Whitney Museum of American Art at Altria, the images that I called *Monochrome Landscapes* were designed to converse formally with the work of another artist in the Whitney collection, Ellsworth Kelly. His monochromes were *Green*, *Black*, *Red*, and *Blue*, and screenprinted with ink.[48] My monochromes are *White*, *Blue*, *Green*, and *Yellow*. They are Cibachrome prints, based on digital files, displayed as forty-by-eighty-two-inch panels on the museum wall. They show four spots on Earth, captured by Ikonos and QuickBird at high resolution, in which almost nothing but snow, water, trees, and sand is visible. Now they are photographs: information, surface, pattern, chance encounter, event, memory, field of color.

There is more than a formal aesthetic at stake here — these are vulnerable ecologies and politically charged landscapes. My insistence on representing them as art and from above, at a distance, was a choice; the resolution of the pixels and the networks of knowledge I used to select the scenes resulted in some surprising discoveries. I was interested in the idea that the places on Earth that appeared from above as more or less a single color were also places that were contested, fragile, and subjected to an increasingly thorough surveillance apparatus.

The white image, of the Arctic National Wildlife Reserve (ANWR) in Alaska, was in the news at the time. The debate over opening this protected space to oil drilling coincided with the larger debates surrounding the Bush administration's plan to go to war in Iraq.

The blue of the Atlantic — with the sensors aimed at the spot where the Greenwich meridian intersects the equator, the "zero-zero" point of latitude and

Ellsworth Kelly, *Four Panels*, 1970–71. © ELLSWORTH KELLY AND GEMINI G.E.L., LOS ANGELES

longitude—attracted me as a sort of digital analog to the abstraction of the monochrome.

Green in this exhibit marked a turning point, my first collaboration with an NGO on an investigative project using satellite imagery. Working with Global Forest Watch, an old-growth tropical rain forest in Yaounde, Cameroon, became the focus of an intensive exploration of illegal logging.

It was the battlefield of that war that other, more obvious, war, in Iraq—then in its second week—that was captured in the yellow image of the desert. For a certain time at the beginning of the fighting, the desert was overexposed to cameras and publicity. Later, images became more difficult to obtain.

This work was extended into a new project. *Shades of Green*, first produced for the group exhibition *Antiphotojournalism* at La Virreina Centre de la Imatge in Barcelona. The show explored new relations between journalism and imagery, and I chose to continue the project begun in *Monochrome Landscapes* with more images of tropical forests, this time underlining the multiple shades of green in the "same" forests.

MONOCHROME LANDSCAPES

New York, 2004—I wanted a series of monochrome landscapes, and so I asked for pictures of places on Earth primarily characterized by one of four basic colors: white, blue, green, and yellow. The rules were simple, and generically, there were not many choices: snow, water, trees, and sand. The satellites had been looking at the Arctic National Wildlife Refuge in Alaska and the Southern Desert in Iraq, and they had to go over the middle of the Atlantic Ocean and the Cameroonian rain forest. For each of these places, I purchased the image data corresponding to an eight-by-eight-kilometer square (about five by five miles) of the Earth's surface. Two were already in the DigitalGlobe archives (Alaska and Iraq), and two required new tasking (QuickBird at 0.61-meter resolution over the Atlantic, Ikonos at 1-meter resolution over Cameroon). The results evoke questions that are at once aesthetic and geopolitical, mapping some of the most vulnerable landscapes of our time—the sparsely populated, resource-rich other sides of globalization.

ZOOM

Over a 16.5-by-16.5–kilometer surface—the footprint of a full QuickBird image—a progressive zoom is already implied, right down to the two-thirds of a square meter that constitutes the pixel itself. But the monochromatic character of the images, in which every pixel looks pretty much alike, makes looking for something in particular rather complicated. However, it's hard *not* to look for something—even if it's just the pixels themselves.

On March 20, 2003, the same day that Operation Iraqi Freedom began, the United States Senate voted against allowing oil drilling in Alaska, despite the fact that it was a priority for President Bush. I was happy about the result of the vote, but the coincidence of the two events pushed me to look for an image of the Arctic National Wildlife Reserve.[49] Geologists estimate that the Alaskan Coastal Plain area in ANWR harbors about 10.4 billion barrels of recoverable crude oil, as well as polar bears, musk oxen, caribou, and scores of migratory bird species. Parts of what is now ANWR have been federally protected since the Eisenhower administration in 1960, and the present refuge, the largest protected wilderness area in the United States at about 19 million acres, dates from the Alaska National Interest Lands Conservation Act (ANILCA) of 1980. Its status continues to be a source of contention in the U.S. House and Senate, even now.

No part is more contentious that the so-called area 1002, which was set aside in the law creating ANWR for further study. According to a U.S. Geological Survey fact sheet, "in section 1002 of [the ANILCA], Congress deferred a decision regarding future management of the 1.5-million-acre coastal plain ('1002 area') in recognition of the area's potentially enormous oil and gas resources and its importance as wildlife habitat."[50] The species of primary concern is the porcupine caribou. (The area is also home to about two hundred and fifty indigenous Inupiat people in the village of Kaktovik.) Throughout the 1990s and 2000s, the question of drilling in area 1002 has been a regular source of debate and controversy in U.S. politics.

BLUE

In October 1884, representatives of twenty-five nations met at the International Meridian Conference in Washington to "fix upon a meridian proper to be employed as a common zero of longitude and standard of time-reckoning throughout the whole world."[51] The zero latitude had been established long before by way of astronomy. The history of the zero-zero point is really the history of modern cartography, which means a history of the representation of the world on an abstract grid. This grid underlies multiple representational systems, different ways to coregister mapping coordinates with physical space. Each representational system has its own politics and history. The Mercator projection prioritizes navigation, while the Peters projection favors what is often called a more politically accurate representation of the Southern Hemisphere, which appears larger than the Northern Hemisphere.[52]

Digital cartography relies on the same systems of alignment, measuring space in two dimensions from the zero-zero point. From environmental calculations, to the selection of military targets, to the timekeeping of the atomic clock, much of our relation to physical space relies on measurements from the zero-zero point in digital space. But it is rarely displayed as a space with physical features. The satellite found it, south of Ghana and west of Gabon, in the Atlantic Ocean.

When the makers of Natural Earth, a public-domain map data set, released the guidelines for version 1.3 of its free raster and vector data to cartographers and mapmakers worldwide, in January 2011, the data came with a caveat:

> WARNING: A troubleshooting country has been added with an Indeterminate sovereignty class called Null Island. It is a fictional, 1 meter square island located off Africa where the equator and prime meridian cross. Being centered at 0,0 (zero latitude, zero longitude) it is useful for flagging geocode failures which are routed to 0,0 by most mapping services.... Null Island in Natural Earth is scale Rank 100, indicating it should never be shown in mapping. Side note: Rank 30 (zoom 29 in Google speak) is 1:1 scale and would require over 288 billion million tiles with a total storage requirement of more than 3.5 billion million megabytes which verges on Borges's essay "On Exactitude in Science." Null Island should only be used during analysis and will keep errant points off your maps.[53]

Borges's story asks about the possibility of making a 1:1 map of the world. Imagine what a model of the world would look like zoomed to three and one-third times its own size.

GREEN

 Guided by the Washington-based NGO Global Forest Watch, an environmental organization that is part of the World Resources Institute, and using a Landsat image as a reference, I ordered a detailed image of a small section of a forest in Cameroon from Ikonos and became a sort of investigator on their behalf.[54] A road is visible. First identified by Global Forest Watch in early 2001, the illegal logging road traverses a not-yet-allocated forest concession area known as UFA (Unité Forestière d'Aménagement) 10-030.[55] The destruction caused by illegal logging in the world's few remaining old-growth tropical rain forests has a direct impact on global biodiversity, climate change, and indigenous people.

For my purposes, this became one of a series of images of vulnerable ecologies, a cautionary image, displaying what happens at the more sparsely populated other side of the globalization upon which many of us depend for resources and carbon absorption. The image has a simple aesthetic—a detailed and undulating green forest, seen from above, whose beauty is interrupted by a road that looks almost natural, simply a part of the landscape. But it is new, not natural, and demands that a viewer ask questions about it.

The image relies on a vast global network in the sky and on the ground. Global Forest Watch's logo at the time reminded viewers that "the forests are online." They worked in physical and digital space simultaneously, using satellites to view from a distance (the crucial forests are in places that are very difficult to get to) and eyes on the ground, the local networks that are equally essential for monitoring forests.

For Global Forest Watch, this was the first 1-meter-resolution satellite image they had used, more than eight hundred times as detailed as the Landsat picture. They used it in court as evidence of illegal logging.[56] This area did finally become a legal concession area in 2002 and abides by the rules of sustainable forestry.[57] The World Resources Institute regularly adds new roads to their forest atlas when they become visible in satellite imagery.[58]

YELLOW

 On the day that I was searching the QuickBird archive for images of Iraq around March 30, the idea of the monochrome landscapes had not yet completely been formed as the structure of this show. I was much more interested in following the news of Operation Iraqi Freedom and purchasing satellite images of Baghdad. In particular, I was interested in the day the museums had been looted, and I wanted to see where United States tanks were located in the city that day. Large armored vehicles would be easy to recognize on a 61-centimeter-per-pixel-resolution image. Although no shutter control had been exercised by the U.S. government, DigitalGlobe was informally and voluntarily "not distributing" imagery of Baghdad in which U.S. troops might be visible.

So I looked around for an image they *would* sell me, and my criteria became rather minimal: I was willing to take an image from any day in the first two weeks of the war that was in the archive and that they would sell to me that day. All I wanted was a document of the war. When I received the image on CD in the mail a few days later, I zoomed in on the monochrome desert and suddenly noticed something to focus on—two helicopters flying in formation, during the second week of Operation Iraqi Freedom, somewhere between Al Busayyah and An Nasiriyah.

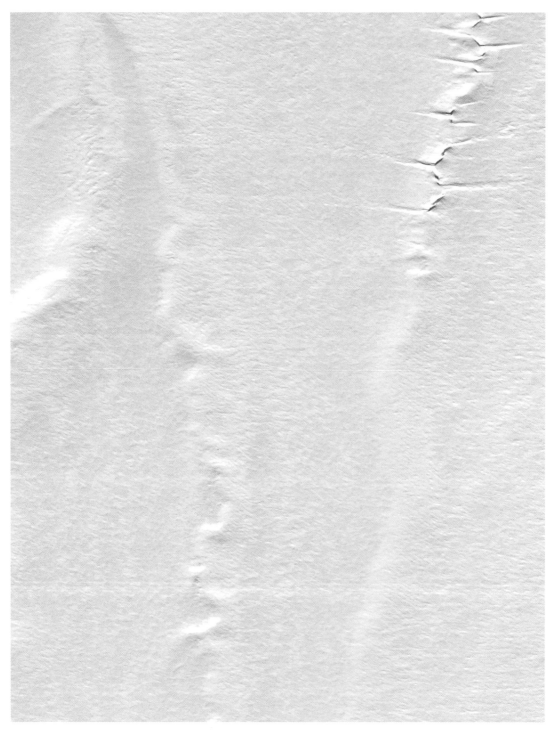

WHITE Area 1002, Arctic National Wildlife Refuge, near Kaktovik, Alaska: zoom 1

BLUE Atlantic Ocean, intersection of Equator/prime meridian, south of Ghana and west of Gabon: zoom 1

WHITE Area 1002, Arctic National Wildlife Refuge, near Kaktovik, Alaska: zoom 2

BLUE Atlantic Ocean, intersection of Equator/prime meridian, south of Ghana and west of Gabon: zoom 2

WHITE Area 1002, Arctic National Wildlife Refuge, near Kaktovik, Alaska: zoom 3

BLUE Atlantic Ocean, intersection of Equator/prime meridian, south of Ghana and west of Gabon: zoom 3

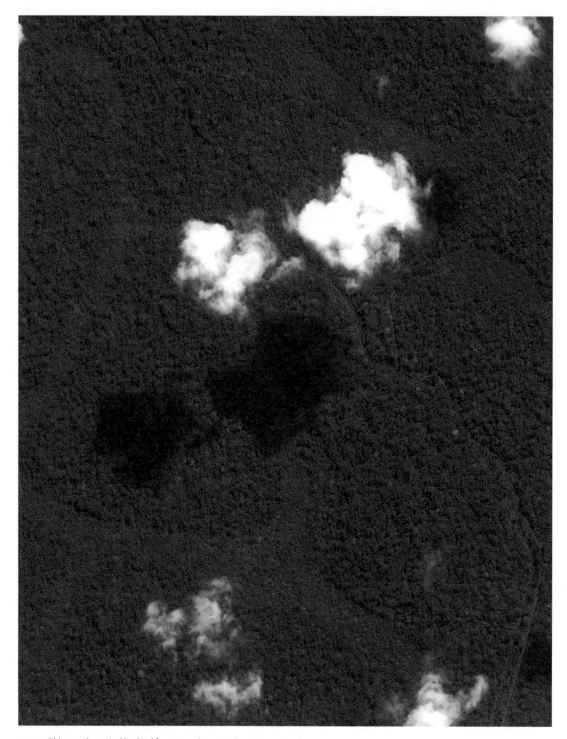

GREEN Old-growth tropical lowland forest, southeastern Cameroon: zoom 1

YELLOW Southern desert, southeastern Iraq, between Al Busayyah and An Nasiriyah: zoom 1

GREEN   Old-growth tropical lowland forest, southeastern Cameroon: zoom 2

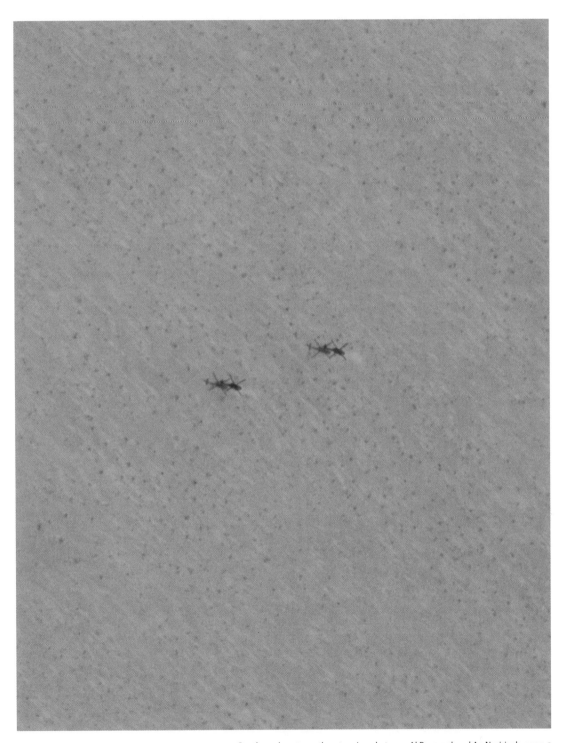

YELLOW Southern desert, southeastern Iraq, between Al Busayyah and An Nasiriyah: zoom 2

GREEN Old-growth tropical lowland forest, southeastern Cameroon: zoom 3

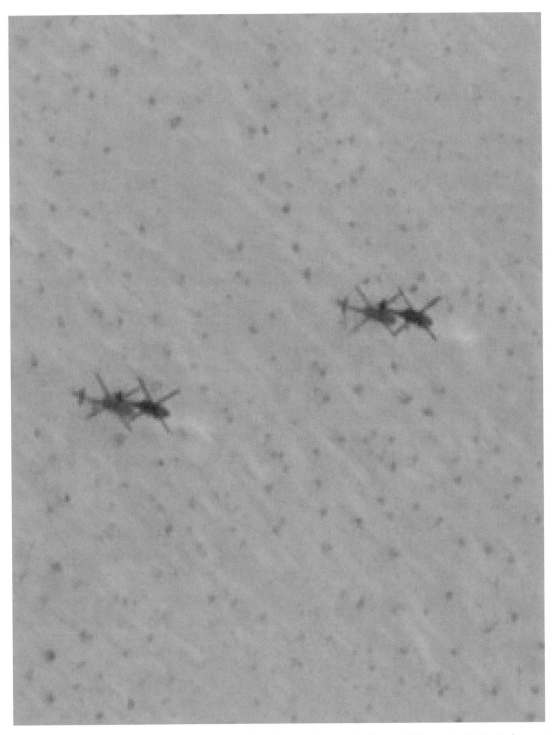

YELLOW Southern desert, southeastern Iraq, between Al Busayyah and An Nasiriyah: zoom 3

Barcelona, 2010 — Shades of Green was produced for the group exhibition *Anti-photojournalism* at La Virreina Centre de la Imatge in Barcelona in 2010. The show examined new modes of producing images of news events. I had become interested in the clearing of old-growth forests by fire, rather than by logging, and I thought it would be possible to continue the project begun with *Monochrome Landscapes* by using satellite images to show the black of the burned rainforest.

I called François-Michel Le Tourneau of the Centre National de la Recherche Scientifique (CNRS), an expert on this, and asked him for some coordinates, which I thought I would give to GeoEye when placing an order. To my surprise, he laughed. Because of the fast rate of growth in the Amazon, he said, I would have had to know the exact time and date of the fire. After the forest is burned, the vegetation grows back very quickly.

So, deprived of black, I returned to green. Building on the color scheme implied by Le Tourneau's analysis, I decided to highlight the shades of green that are now the hallmark of vulnerable rainforests — new crops of palms and soybeans, in contrast to the surrounding trees and mature plant life that they quickly deforest and replace.

These images, acquired by satellites in 2008 and 2009, show the transformation of forests in Indonesia and Brazil under the pressure of agriculture, settlement, fire, cattle ranching, and logging. Change, or "forest cover loss," is registered in different shades of green: dark green for original forest, green for recently burned and depleted forest, and light and very light green for pastures or crops.

According to the *2010 Global Forest Resources Assessment* from the Food and Agriculture Organization of the United Nations, deforestation — the conversion of tropical forests into agricultural land — shows signs of decreasing in several countries, but continues at a high rate in others. Around 13 million hectares of forest (32 million acres) were converted to other uses or lost through natural causes each year in the last decade, compared with 16 million hectares per year (just under 40 million acres) in the 1990s. Both Brazil and Indonesia, which had the highest net loss of forest in the 1990s, have significantly reduced their rate of loss today. Nevertheless, forests continue to disappear.

SHADES OF GREEN

Itauba County, state of Mato Grosso, Brazil. Itauba
County covers about 4,500 square kilometers, just
south of the Amazon, with a population of less
than 5,000 people. Agricultural lands now extend
well into former forests, primarily for cattle ranching
and soybean farming.

Acquired: June 28, 2008, 21:12:49 GMT
Upper left coordinates:
11°9′25.92″ W, 55°31′6.24″ S
Lower right coordinates:
11°9′17.64″ W, 55°24′41.76″ S
Ikonos satellite, 1 m per pixel

SHADES OF GREEN

Plantations are estimated to have grown from
106,000 hectares in 1960 to 18 million hectares in
2006. Pictured here are logging roads in West
Kalimantan in 2009, making one of the few
remaining pristine tropical forests in Indonesia
vulnerable to deforestation.

Acquired: February 16, 2009, 03:08:00 GMT
Upper left coordinates:
1°39′1.67″ S, 110°43′0.75″ W
Lower right coordinates:
1°38′57.79″ N, 110°51′55.18″ W
GeoEye-1 satellite, 0.5 m per pixel

SHADES OF GREEN Itauba County, state of Mato Grosso, Brazil: zoom 1

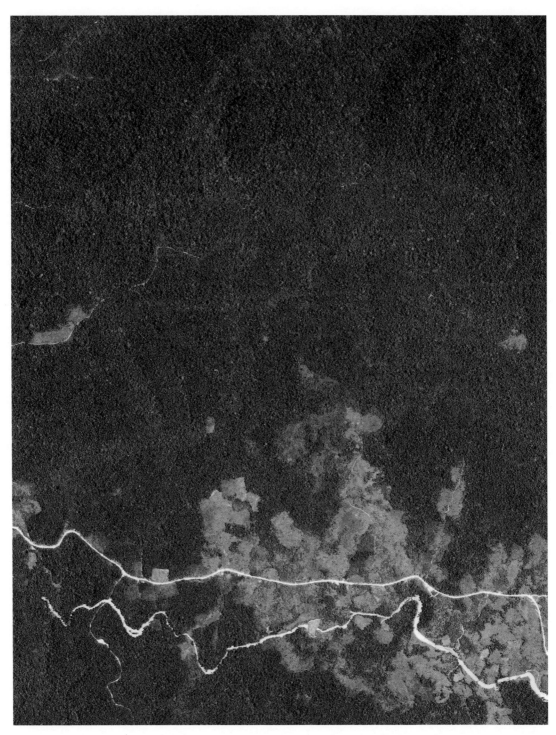

SHADES OF GREEN Kalimantan, Indonesia, border of Kalimantan Barat and Kalimantan Tengah: zoom 1

SHADES OF GREEN Itauba County, state of Mato Grosso, Brazil: zoom 2

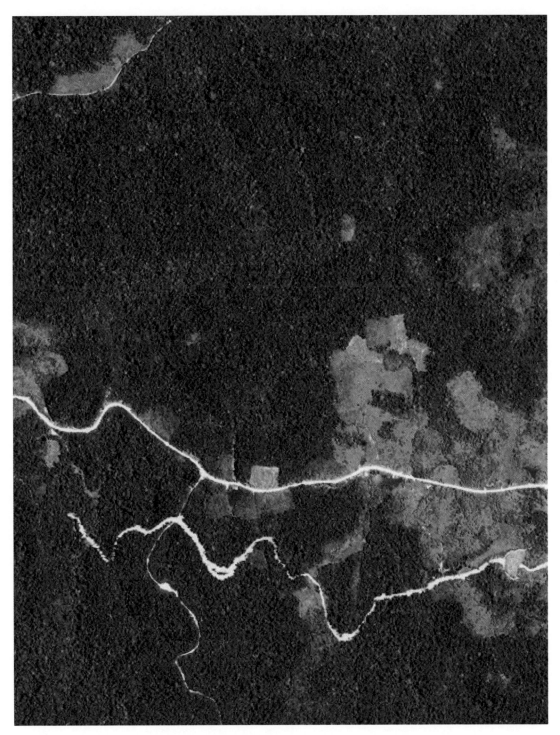

SHADES OF GREEN Kalimantan, Indonesia, border of Kalimantan Barat and Kalimantan Tengah: zoom 2

SHADES OF GREEN Itauba County, state of Mato Grosso, Brazil: zoom 3

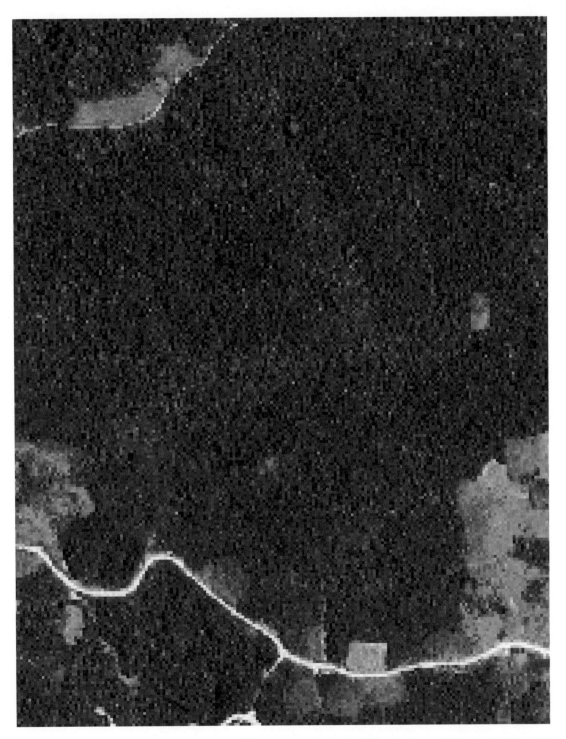

SHADES OF GREEN Kalimantan, Indonesia, border of Kalimantan Barat and Kalimantan Tengah: zoom 3

Daten des globalen Devisenhandels:
Euro und Yen in Dollar angegeben,
30 Dezember 1999, gemeldet von Reuters.
Die intensivsten fuenf Minuten
steuern die Uhrzeiger.

Worldwide foreign exchange data:
euro and yen expressed in dollars,
30 December 1999, as reported by Reuters.
The densest five minutes are driving
the hands of the clock.

Daten des globalen Devisenhandels:
Euro und Yen in Dollar angegeben,
31 Dezember 1999 bis **1. Januar 2000**,
gemeldet von Reuters.
Die Daten aus 30 Stunden
sind in fuenf Minuten dargestellt.

Worldwide foreign exchange data:
euro and yen expressed in dollars,
31 December 1999 - **1 January 2000**,
as reported by Reuters.
30 hours of data are expressed in five minutes.

Daten des globalen Devisenhandels:
Euro und Yen in Dollar angegeben,
3 Januar 2000, gemeldet von Reuters.
Die intensivsten fuenf Minuten
steuern die Uhrzeiger.

Worldwide foreign exchange data:
euro and yen expressed in dollars,
3 January 2000, as reported by Reuters.
The densest five minutes are driving
the hands of the clock.

8 Global Clock

Nothing happened

New York, 2012 — As the turn of the millennium approached, I had an interesting opportunity to take my work in a new direction. I had obviously done a lot of work with data, but in general, it was visual data, maps or GPS readouts or satellite imagery. I realized that it might be just as interesting to work with data that had a spatial component, but was not already visual, which is to say to explore what is now known as "data visualization."

I was interested in movement, in flows of data across borders and around the world, especially very high-speed flows, and in the challenges that some of these flows posed to the way we think about space. I was invited to participate in an exhibition in Düsseldorf organized around the question of money (it was called *The Fifth Element*), scheduled to open in January 2000. I had to plan for and install the exhibit in 1999, and so I decided to record and archive financial data transactions across the turn of the millennium. You can probably remember the panic about what was nicknamed "Y2K" and the fear that computers and the networks that linked them would somehow collapse as their software encountered dates that began with 20 rather than 19. I thought it would be interesting to track the flow of data about money across the millennium shift.

I also wanted to make a point about a turn that was taking place in architectural discourse at the time, one that polemically positioned itself "against critique," or rather, "postcritically," in favor of what was called "propositions." The desire to get beyond critique felt to me like a massive capitulation to a certain form of the status quo, particularly a global financial one, and most especially a refusal to investigate the new landscapes of data and technology with anything other than a would-be instrumental relation.[59]

It was a moment that Rem Koolhaas had called the YES regime—the image was ubiquitous in his lectures of the time, and it caught the eye of a journalist for *Wired* magazine. In a lecture at U.C. Berkeley, Koolhaas outlined plans for an expanded Schiphol Airport, outside Amsterdam. The journalist paraphrases the presentation—"The airport-island will be funded and supported by commercial enterprises. It will be a shopping heaven."—and then comments:

> To ensure that even the most innocent among them got the point, he put up a slide showing the symbols for the yen, the euro, and the dollar. "During some recent work at OMA," he said, "we noticed that the signs of the world's major currencies, put together, spell YES. We are working inside this global YES." [60]

And Koolhaas, to his credit, was not shy about drawing out the most radical implications of this "working inside." In another lecture around the same time, referring once again to the three currencies, he said:

> They describe a regime under which we are all active and willing. On the one hand, it is a regime that sets our parameters, and those parameters are fairly immutable. But on the other hand, it is also a regime that gives us an almost unbelievable amount of freedom to establish our own trajectories within it, and also to establish whatever connections within it, including connections between not only people but between different enterprises.[61]

Postcriticality had everything to do with money, with saying YES to money, in whatever currency, with not thinking much about money besides how to make it. I wanted to do something else with the currencies.

To be honest, I expected the computers to crash as I was recording the data over the millennium and that that my show would be a replay of the event—whatever form that took. But instead, as I watched the celebrations that New Year's Eve all around the world, starting in Australia, on TV and on the Internet, and I realized my installation was doing just fine in Düsseldorf.

Nothing happened.

GLOBAL CLOCK

Düsseldorf, 1999–2000—This installation was exhibited in *The Art of Money* at the Kunsthalle Düsseldorf in 1999/2000 and later in *Money and Value*, Biel/Bienne, EXPO.02, Swiss National Exposition. The first version used recorded financial data feeds from Reuters to track the movements of these three currencies in the days and hours immediately surrounding the turnover of the millennium. The later version displayed the data in real time over the duration of the exhibition. When trading stopped, so did the clock.

Money is, among many other things, a way of exchanging things that aren't similar. Money can also change hands faster than things. Rather than simply standing in for things being exchanged, sometimes money itself, as such, is exchanged.

The installation represents on a clock the relative values of the euro, the yen, and the dollar as they were reported moment by moment at the turn of the century by the Reuters global financial data service. Together, these three currencies account for approximately 90 percent of the global foreign exchange market.

Today, in a networked world, money moves from place to place as data, invisibly, across cables and satellites at the speed of light. Here, a feed from a global financial information network is rerouted into the museum—not so much to render the invisible visible, but to investigate the luminous immateriality of money and its mutable media. The medium of this money is light, and it shows up periodically on screens as the translation of invisible into visible light. Here, two interfaces with foreign-exchange data provided by Reuters are displayed on the walls. In the foreign-exchange market, where roughly 1.5 trillion U.S. dollars move around the globe every day, the relative value of currencies fluctuates over time at high speeds and short intervals.

Money, the storehouse of value, changes its value from second to second. Typically, these changes in value are represented as the jagged line of a graph that moves over time: the standard Reuters real-time interface does just this. *Global Clock* builds a new interface for seeing financial data in motion. The three hands of a clock move in accordance with the irregular rhythms of the foreign-exchange market. The change in values of the euro and the yen are each plotted against the dollar and presented as a moving dot along the straight lines of the green (hour) hand and the yellow (minute) hand of the clock. In these data feeds, the red dollar (second hand) is the invisible standard against which the euro and the yen are measured. As their dollar valuations change, the euro and yen symbols slide along their respective hands; meanwhile, the dollar remains stationary on the red hand.

The high-speed flow of foreign exchange was recorded over the course of the exhibition, which spanned the turn of the millennium. While the world was watching the celebrations make their way around the globe, following the path of latitudes across the longitudinal lines, the clock just kept on ticking. There had been ominous predictions that computer systems would fail, confused by the disappearance of the digits "19" in the electronic date field. Nothing failed. I have clipped a few images from the densest five minutes of currency trading on the last business day of 1999 (December 30, 9:25 to 9.30 a.m.) and again the densest five minutes on the first full business day of 2000 (January 3, 4:55 to 5:00 p.m.), and in between, the entire data flow of December 31, 1999 and January 1, 2000, compressed into a very fast five minutes of transactions were recorded by Reuters, worldwide, and nothing much happened. The clock just kept ticking, and the money just kept moving.

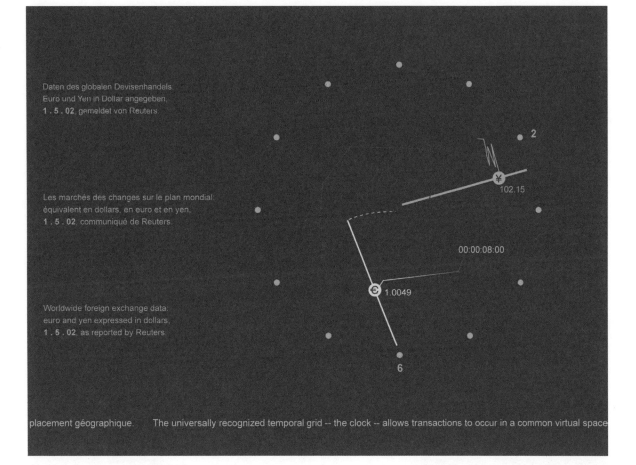

Daten des globalen Devisenhandels:
Euro und Yen in Dollar angegeben,
1 . 5 . 02, gemeldet von Reuters.

Les marchés des changes sur le plan mondial:
équivalent en dollars, en euro et en yen,
1 . 5 . 02, communiqué de Reuters.

Worldwide foreign exchange data:
euro and yen expressed in dollars,
1 . 5 . 02, as reported by Reuters.

2

¥
102.15

00:00:08:00

€ 1.0049

6

placement géographique.   The universally recognized temporal grid -- the clock -- allows transactions to occur in a common virtual space

Global Clock No. 2, driven with Reuters data in real time, foregrounding time, value, and volume of currency exchanges. The dollar activates the second hand, the euro the minute hand, and the yen the hour hand.

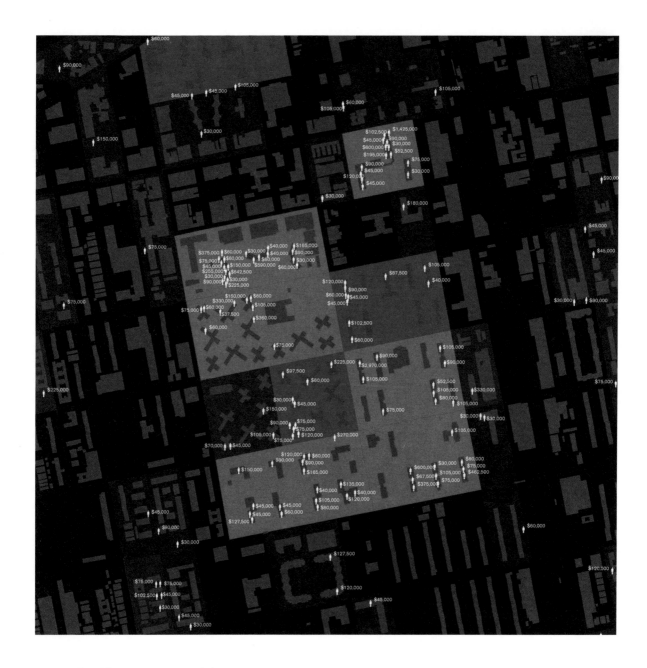

It cost 17 million dollars to imprison 109 people from
these seventeen blocks of Brownsville, Brooklyn,
New York, in 2003.

9 Million-Dollar Blocks

The "most phenomenal" fact of all

New York, 2012 — Eddie Ellis spent twenty years in prison in New York State. In 1992, just after he had been released and returned to his home in Harlem, he told *New York Times* reporter Francis X. Clines about research that he and other prisoners had done while incarcerated, research in what the story called "a prisoner's 'think tank' at Green Haven prison in Stormville, N.Y."[62] Of all that they had learned, a pair of "hard facts" stood out, he said: "the fact that more than 85 percent of prisoners in the state are black or Latino and — most phenomenal of all — that 75 percent of the state's entire prison population comes from just seven neighborhoods in New York City." The article, which ran on the front page of the paper two days before Christmas that year, went on to explain that this second fact, "that three out of four prisoners come from, prey upon and return to seven neighborhoods encompassed by just 18 of the state's 150 Assembly districts, or 12 percent of the population, is at the heart of Mr. Ellis's new mission as an unaccredited street penologist without portfolio." The story was accompanied by a map, the caption of which read: "Map of New York City, indicating seven neighborhoods where three out of four New York State prisoners come from."

Ellis's home-grown research mission — and in particular, the map — caught the eye of other scholars and advocates for criminal justice reform. A year later, Lola Odubekun published the *Vera Institute Atlas of Crime and Criminal Justice in New York City*, which, in addition to its rather predictable crime maps, also included two maps of incarceration: one titled "Rikers Island Inmates by Home Residence, March 1993" and another titled "Distribution of Persons Arrested by Neighborhood of Residence, 1989."[63] Although the report noted that "69 percent of the 64,501 inmates in the state prisons were from New York City,"[64] and although the maps clearly showed that the vast number of those inmates came from very few neighborhoods in the city, no conclusions were drawn noting the unusual statistical concentration.

Five years later, Eric Cadora of the Center for Alternative Sentencing and Employment Services made the decisive move to begin acquiring data about incarceration from state criminal justice records themselves in order at once to test these early cartographic projects at a larger scale and to draw some conclusions: to show that incarceration is a problem of the city and to demonstrate that policy needed to address the issue directly. He called the project "justice mapping." Cadora, working with Charles Schwarz, produced a different sort of map, one that, as he told Jennifer Gonnerman in the *Village Voice*, "would help people envision solutions rather than just critiques."[65] As Gonnerman reported, "they made a series of maps illustrating where inmates come from and how much money is spent to imprison them," and there they discovered what came to be called "million-dollar blocks."

In 2005, a study of million-dollar blocks became the first project of the Spatial Information Design Lab (SIDL), which I had started the year before at the Graduate School of Architecture, Planning, and Preservation at Columbia University. Over a number of years and in a variety of different ways, with dozens of maps of neighborhoods across the United States, the research built on Cadora's project and took up the challenge of making visible a decidedly spatial phenomenon, but one that still remained difficult to see.

One reason for the difficulty is that the geography of incarceration is both a micro and a macro feature of contemporary urbanism. Looking at the block is essential, but it fails to make much sense unless it's seen within the context of a larger metropolitan infrastructure of criminal justice and social services...and vice versa.

To show this, *Million-Dollar Blocks* borrows and inverts the language of crime "hot spot" maps. Introduced by New York City police commissioner William Bratton in 1994 with the enthusiastic endorsement of Mayor Rudolph Giuliani, the COMPSTAT ("computerized statistics") program used GIS software to map the locations and times of crimes across New York City.

Million-Dollar Blocks shifts the frame ever so slightly and makes use of otherwise rarely accessible data, also collected by the criminal justice system, to corroborate Ellis's early research. Simply by mapping the home addresses of people as they are admitted to prison, which are also the addresses to which they will most likely return upon release, and by correlating that with the amount of time they spend in prison (and hence the cost to the state), "phenomenal facts" indeed emerge.

The maps show the disproportionate concentrations of incarceration in poor and isolated city blocks across the United States. The project aggregates data and then zooms in to the microgeographies of those communities, mining existing data and repurposing it to produce new visual and quantitative meanings. In so doing, the maps direct viewers to look more closely at certain places, for instance, the Brooklyn neighborhood of Brownsville, and ask: "What's behind the red polygon?"

New York, 2006 — The United States currently has more than two million people locked up in jails and prisons. A disproportionate number of them come from a very few neighborhoods in the country's biggest cities. In many places, the concentration is so dense that states are spending in excess of a million dollars per year to incarcerate the residents of single city blocks. When these people are released and reenter their communities, roughly 40 percent do not stay more than three years before they are reincarcerated.

Using rarely accessible data from the criminal justice system, the Spatial Information Design Lab and the Justice Mapping Center have created maps of these "million-dollar blocks" and the city-prison-city-prison migration flow for five of the nation's cities. The maps suggest that the criminal justice system has become the predominant government institution in these communities and that public investment in this system has resulted in significant costs to other elements of our civic infrastructure: education, housing, health, and family. Prisons and jails form the distant exostructure of many American cities today.

Have prisons and jails become the mass housing of our time? How has the War on Drugs affected incarceration rates? What are the differences between crime maps and prison admission maps? What are the relationships between prison populations and poor communities? Has incarceration become a response to poverty, rather than to crime? What are the relationships between jailed populations and homeless ones?

The relationships implied by these questions become evident when criminal justice data is aggregated geographically and visualized in maps. The focus shifts away from a case-by-case analysis of the crime and punishment of an individual, away from the geographic notation of crime events, and toward a geography of incarceration and return.

The maps pose difficult ethical and political questions for policy makers and policy designers. When they are linked to other urban, social, and economic indicators of incarceration, they also suggest new strategies for approaching urban design and criminal justice reform together.

WHY ARE SO MANY AMERICANS IN JAIL AND PRISON?

Since 1970, Americans have been living in an era of what some have called mass incarceration, one of the "greatest social experiments of our time."[66] The crime rate in America over the course of the last century has moved up and down in a periodic wave. The corresponding rates at which Americans have been incarcerated look very different. In contrast to the periodic undulations of the crime rate, the incarceration rate remained constant for most of the century.

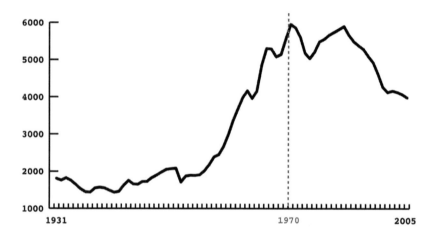

Crime rates form a relatively self-consistent wave of activity.

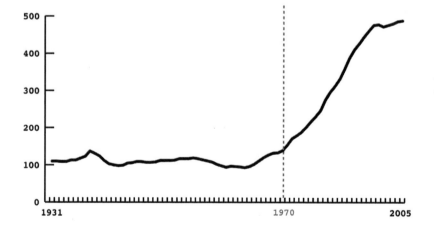

Incarceration rates remain relatively constant until 1970, when a radical upward trend is driven by policy.

From the late 1970s, however, it has been climbing rapidly. The result has been a tenfold increase in the standing prison population, from two hundred thousand in 1970 to two million in 2000.[67] How we respond to crime is a matter of values, decisions, and policy, all the way down to the basic questions defining what counts as a crime. In the late 1970s and early 1980s, efforts to fight poverty were systematically replaced by the War on Drugs, including the criminalization of most drug offenses. Crime became the surrogate for poverty and incarceration the primary response.

Poverty policy in the United States since 1900.

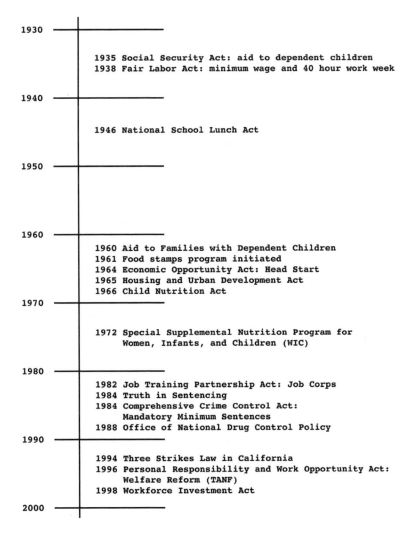

1930

1935 Social Security Act: aid to dependent children
1938 Fair Labor Act: minimum wage and 40 hour work week

1940

1946 National School Lunch Act

1950

1960

1960 Aid to Families with Dependent Children
1961 Food stamps program initiated
1964 Economic Opportunity Act: Head Start
1965 Housing and Urban Development Act
1966 Child Nutrition Act

1970

1972 Special Supplemental Nutrition Program for
     Women, Infants, and Children (WIC)

1980

1982 Job Training Partnership Act: Job Corps
1984 Truth in Sentencing
1984 Comprehensive Crime Control Act:
     Mandatory Minimum Sentences
1988 Office of National Drug Control Policy

1990

1994 Three Strikes Law in California
1996 Personal Responsibility and Work Opportunity Act:
     Welfare Reform (TANF)
1998 Workforce Investment Act

2000

FROM DATA TO MAPS

A criminal justice data set is most commonly maintained and presented as a list. It is designed to track people as individual cases. As individuals make their way through the system, information is entered into a database and accumulates: name, crime, length of sentence, home address, and so on. Individually, the information forms a portrait of a case. Aggregated, the cases create a statistical portrait of a society.

When maps are made from data such as these, they often stop at the very first element: what crimes were committed and where. Crime maps have played a significant role in the public discourse on cities over the last thirty years. These maps have, in fact, become one of the most prominent instruments through which we understand and interpret our cities.

```
30-Jun   3:09 A.M.   HM445540   3300 BLOCK W. 23RD ST.          SIDEWALK
26-May   1:10 A.M.   HM373655   2600 BLOCK S. TRUMBULL AVE.     STREET
11-May   11:30 P.M   HM345311   1000 BLOCK N. MONTICELLO AVE.   SIDEWALK
4-Jul    11:48 P.M.  HM455004   5400 BLOCK W. WRIGHTWOOD AVE.   APARTMENT
1-Jul    4:45 P.M    HM448717   1500 BLOCK W. 77TH ST.          APARTMENT
24-Jun   12:54 A.M.  HM433672   13400 BLOCK S. BALTIMORE AVE.   RESIDENCE
23-Jun   9:23 A.M.   HM431903   4900 BLOCK W. WALTON ST.        GROCERY FOOD STORE
22-Jun   9:47 A.M.   HM429882   8000 BLOCK S. INDIANA AVE.      RESIDENCE
18-Jun   2:40 A.M.   HM421205   6200 BLOCK S. KIMBARK AVE.      RESIDENCE
16-Jun   10:08 A.M.  HM417358   4700 BLOCK N. SPAULDING AVE.    APARTMENT
11-Jun   8:30 P.M.   HM408837   1100 BLOCK W. JACKSON BLVD.     RESIDENCE
2-Jun    3:28 A.M.   HM388296   11600 BLOCK S. ASHLAND AVE.     RESIDENCE
31-May   1:37 A.M.   HM383920   2900 BLOCK N. SHERIDAN RD.      RESIDENCE
30-May   2:30 P.M.   HM382758   3100 BLOCK W. BYRON ST.         SCHOOL BUILDING (PUBLIC)
29-May   2:48 A.M.   HM379750   2100 BLOCK S. HARDING AVE.      RESIDENCE: PORCH/HALLWAY
29-May   12:30 A.M.  HM379786   3900 BLOCK W. MADISON ST.       SMALL RETAIL STORE
23-May   6:45 A.M.   HM367460   6600 BLOCK S. RHODES AVE.       RESIDENCE: PORCH/HALLWAY
20-May   4:14 A.M.   HM361811   900 BLOCK W. 52ND ST.           APARTMENT
19-May   9:37 A.M.   HM359905   1100 BLOCK W. 110TH ST.         RESIDENCE
16-May   8:43 P.M.   HM355107   3900 BLOCK W. DIVERSEY AVE.     APARTMENT
14-May   5:10 A.M.   HM349864   3800 BLOCK W. NORTH AVE.        RESIDENCE
11-May   2:07 A.M.   HM343234   4300 BLOCK W. WILCOX ST.        RESIDENCE: PORCH/HALLWAY
8-May    5:41 A.M.   HM337007   6200 BLOCK N. SHERIDAN RD.      RESIDENCE
4-Jul    10:13 P.M.  HM454861   5400 BLOCK W. GRACE ST.         SIDEWALK
4-Jul    6:34 P.M.   HM454560   3900 BLOCK W. MADISON ST.       RESIDENCE
1-Jul    1 P.M.      HM454078   5600 BLOCK S. NASHVILLE AVE.    RESIDENCE
11-Mar   3:50 A.M.   HM225590   6300 BLOCK S. MORGAN ST.        VEHICLE: NON-COMMERCIAL
4-Jul    10:40 P.M.  HM454929   2100 BLOCK W. CULLERTON ST.     RESIDENCE: GARAGE
3-Jul    8:02 A.M.   HM451516   2900 BLOCK N. SHERIDAN RD.      RESIDENCE: PORCH/HALLWAY
3-Jul    12:35 A.M.  HM451157   1800 BLOCK W. 51ST ST.          RESIDENCE
2-Jul    11:45 P.M.  HM451279   6300 BLOCK S. MORGAN ST.        COMMERCIAL / BUSINESS OFFICE
2-Jul    10:59 P.M.  HM451096   9200 BLOCK S. DAUPHIN AVE.      STREET
2-Jul    8:30 A.M.   HM449799   3200 BLOCK W. WARNER AVE.       VEHICLE: NON-COMMERCIAL
1-Jul    3:13 P.M.   HM448484   1200 BLOCK N. SPRINGFIELD AVE.  STREET
1-Jul    4:20 A.M.   HM447631   7300 BLOCK S. HONORE ST.        VEHICLE: NON-COMMERCIAL
30-Jun   7:41 A.M.   HM445573   1700 BLOCK W. 47TH ST.          RESTAURANT
29-Jun   1:47 A.M.   HM443456   6600 BLOCK S. CARPENTER ST.     STREET
28-Jun   10:05 P.M.  HM443218   11100 BLOCK S. WENTWORTH AVE.   VEHICLE: NON-COMMERCIAL
28-Jun   11:07 A.M.  HM441966   1800 BLOCK S. DRAKE AVE.        VEHICLE: NON-COMMERCIAL
28-Jun   1:57 A.M.   HM441411   3000 BLOCK W. CULLERTON ST.     VEHICLE: NON-COMMERCIAL
27-Jun   5:05 A.M.   HM439509   5900 BLOCK S. ALBANY AVE.       VEHICLE: NON-COMMERCIAL
27-Jun   12:54 A.M.  HM439410   3500 BLOCK W. KEATING AVE.      VEHICLE: NON-COMMERCIAL
26-Jun   1:30 P.M.   HM438209   6300 BLOCK S. MORGAN ST.        STREET
26-Jun   11:30 A.M.  HM437995   4700 BLOCK W. MONROE ST.        SIDEWALK
25-Jun   8:12 P.M.   HM436972   1300 BLOCK S. AVERS AVE.        GOVERNMENT BUILDING/PROPERTY
25-Jun   9:51 A.M.   HM436022   1500 BLOCK N. LAWNDALE AVE.     STREET
25-Jun   3:49 A.M.   HM435694   4500 BLOCK S. SPRINGFIELD AVE.  VEHICLE: NON-COMMERCIAL
```

Excerpt from database,
www.chicagocrime.org.

According to the National Institute of Justice (NIJ), "mapping crime can help law enforcement protect citizens more effectively in the areas they serve. Simple maps that display the locations where crimes or concentrations of crimes have occurred can be used to help direct patrols to places they are most needed. Policymakers in police departments might use more complex maps to observe trends in criminal activity."[68]

Mapping the data about the location of crimes has prompted successful campaigns to transform urban policing from a reactive, calls-for-service approach to an active community policing strategy focused on so-called high-crime locations. Crime maps collect individual incidents over time to identify "hot spots," places that can become the focus of intense police—and political—attention. As the NIJ report puts it (candidly, if rather casually): "using maps that help people visualize the geographic aspects of crime, however, is not limited to law enforcement. Mapping can provide specific information on crime and criminal behavior to politicians, the press, and the general public."[69]

Typical crime map, from www.chicagocrime.org.

Criminal events, not people, are mapped to the city.

FROM CRIME MAPS TO ADMISSIONS MAPS

If crime maps succeeded dramatically in mobilizing public opinion—redefining the city as a mosaic of safe and unsafe spaces and forcing the reallocation and targeting of police resources on specific neighborhoods—the gains were short-lived. The resulting crime prevention techniques and the community policing movement in general soon reached the inevitable limits of any purely tactical approach. The city spaces that were targeted became safer, but too often, crime incidents were simply displaced to other locations.

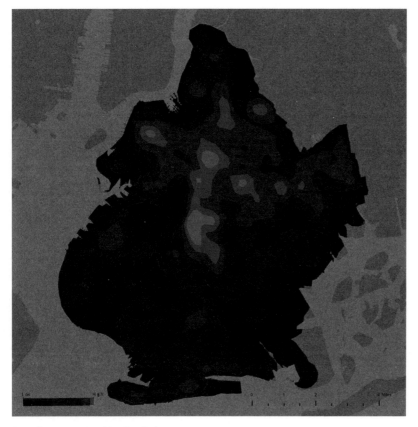

Crime density map, Brooklyn, New York, 1998.

By focusing solely on events, the human underpinnings of crime were left largely unaffected. When we shift the maps' focus from crime events to incarceration events, strikingly different patterns become visible. The geography of prison differs in important ways from the geography of crime. Diffused and dispersed across the city, crime happens in many different places. But the people who are convicted and imprisoned for urban crimes are often quite densely concentrated geographically.

Prison admissions density map, Brooklyn, New York, 2003.

The crime rates in the most affected precincts are typically four times higher than the lowest. But the highest-incarceration-rate precincts show activity upward of ten times higher than those of the lowest-incarceration-rate precincts. Like poverty, incarceration is spatially concentrated, and much more so than crime. It is as if by imprisoning the residents of these neighborhoods—making them disappear from their city—we were simply mirroring the disappearance of the conversation on poverty.

Prison admissions by census tract, Brooklyn, New York, 2003.

Just as the incarceration rate tracks the eclipse of that debate, the geographical inquiry into criminal justice in the city uncovers the territory of the juxtaposition between crime and poverty. Focusing on where incarcerated people live when they are not in prison and comparing that with poverty suggests this conjunction rather starkly. Is incarceration policy the new solution to poverty, or a new structural component?

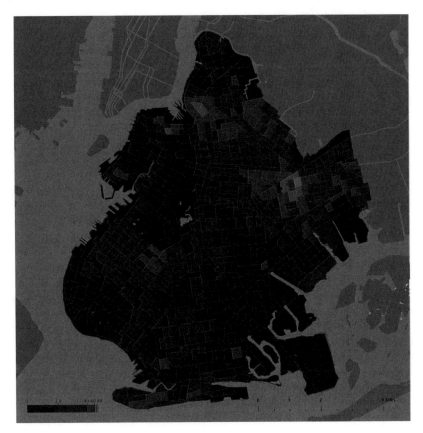

Population living in poverty by census tract, Brooklyn, New York, 2000.

REDEFINING THE PROBLEM: MASS MIGRATION AND REENTRY

Six hundred thousand people return from prison each year in the United States, and millions more come home from jails.[70] About two hundred and forty thousand of the released prisoners—roughly 40 percent—will return to prison within three years.[71] In and out, they come and go, all too often simply cycling back and forth between the same places. New maps can help us grasp this extraordinary phenomenon: prison migration, and with it, high-resettlement communities. When crime maps are replaced by incarceration maps, we can finally visualize the geography of

Prisoner migration patterns, Brooklyn to New York State, 2003.

a massive migration: the flow of people in and out of the city. We can ask whether this quiet but pervasive migration crisis isn't creating a growing class of non-citizens, concentrated in very few places in all of our major cities. The new visualizations reveal what was previously difficult to see—the mass disappearance and reappearance of people in the city. They focus on the systematic phenomenon of ex-prisoners' reentry and examine new institutions that respond to this structural feature of urban life. What happens to these people when they come home? We often know where they are going and what will happen. What is our responsibility to resettle them effectively, given all that we know?

Prisoner migration patterns, Brooklyn, New York, 2003.

04/19/7349 53RD ST	BROOKLYN	NY	49	253	900	300PL	160.15		1F	B	0	5V	ROBBERY-1ST	101	145	06/28/79	23	1New York	BK	349	750000
11/28/8306 MARTENS ST	BROOKLYN	NY	253	1502	300	150PL	160.15		1F	B	0	5V	ROBBERY-1ST	100	140	11/28/83	23	1New York	BK	106	375000
07/11/04016 7 A VENUE	BROOKLYN	NY	153	902	180	90PL	155.42		1F	B	0	8N	GR LARC 1-VALUE > $1 MILLION	100	100	09/10/03	23	1New York	BK	4016	225000
04/23/0377 WILSON AVENU	BROOKLYN	NY	43	23	48	24PL	140.20		3F	D	0	7N	BURGLARY-3RD	100	140	02/25/00	0	1New York	BK	377	60000
01/31/0590 HANC OCK STRE	BROOKLYN	NY	49	33	900	36PL	115.08		1F	B	0	43N	CRIMINAL FACILITATION-1ST	100	140	03/10/77	40	1New York	BK	590	90000
05/30/041 WILSON AVE	BROOKLYN	NY	33	13	36	12VTL	0511	03A2	1F	E	0	99N	AGG UNLIC OPER 1- 10/M[illegible] SUSP	100	100	05/30/03	0	1New York	BK	41	30000
10/02/0793 QUIN CY STREE	BROOKLYN	NY	43	23	48	24PL	140.20		3F	D	0	7N	BURGLARY-3RD	101	140	05/03/85	30	1New York	BK	793	60000
08/12/02960 W 24TH ST	BROOKLYN	NY	33	0	36	0PL	265.02	08	3F	E	1	20V	CRIM POSS WEAP 3RD-AMMO CLIP	100	130	07/07/00	23	1New York	BK	2960	45000
07/13/0135 HEGA MAN ST	BKLYN	NY	153	53	180	60PL	160.15		1F	C	1	5V	ROBBERY-1ST	100	140	02/10/88	0	1New York	BK	135	150000
09/16/9315 LIVONIA AVE	BKLYN	NY	93	542	108	54PL	220.39		3F	B	0	15N	CRIM SALE CONTRL SUBST-3RD	100	140	06/22/94	0	1New York	BK	315	135000
01/07/03773 ATLANTIC AVE	BROOKLYN	NY	49	203	900	240PL	125.25		0F	A	0	1L	MURDER	100	140	06/20/75	0	1New York	BK	1773	600000
08/28/065 GRANI TE ST.	BKLYN	NY	123	43	144	48PL	105.15		2F	B	0	43N	CONSPIRACY -2ND	100	100	06/19/03	23	1New York	BK	65	120000
04/18/0482 ATKI NS AVENU	BROOKLYN	NY	33	13	36	12PL	265.02		3F	D	0	20N	CRIM POSS WEAP-3RD DEG	101	140	12/20/94	0	1New York	BK	482	30000
10/19/0166 TEN EYCKM WK	BKLYN	NY	123	63	144	72PL	220.39		3F	B	0	15N	CRIM SALE CONTRL SUBST-3RD	101	101	04/07/03	23	1New York	BK	166	180000
10/20/9297 1 ST	BROOKLYN	NY	63	33	72	36PL	190.80		1F	D	0	31N	IDENTITY THEFT 1ST DEGREE	100	100	04/07/03	23	1New York	BK	297	90000
08/09/0208 ROCHESTER AV	BROOKLYN	NY	43	23	48	24PL	220.06		5F	D	0	19N	CRIM POSS CONTR SUBST-5TH	101	140	02/14/94	23	1New York	BK	208	60000
05/22/0286 ALABAMA AVEN	BROOKLYN	NY	103	0	120	0PL	265.03		2F	C	0	20V	CRIM POSS WEAPON- 2ND DEGREE	100	100	07/03/03	23	1New York	BK	586	150000
05/28/0305 MALCOLM X BL	BROOKLYN	NY	33	13	36	12PL	220		0	Y	0	0	YO Adm Law Detail Not Known	100	100	08/05/03	62	1New York	BK	105	30000
04/02/0270 GLENMORE AVE	BROOKLYN	NY	83	0	96	0PL	120.10		1F	B	0	6V	ASSAULT 1ST DEGREE	100	100	09/24/03	23	1New York	BK	270	120000
09/21/02535 LINDEN BOULE	BROOKLYN	NY	93	33	108	36PL	220.39		3F	B	0	15N	CRIM SALE CONTRL SUBST-3RD	100	100	08/07/03	0	1New York	BK	2535	90000
08/04/0320 SCHERMERHORN	BROOKLYN	NY	422	0	42	0PL	140.25		2F	C	0	7V	BURGLARY-2ND	100	100	11/07/03	23	1New York	BK	120	52500
04/07/0637 PARK PLACE	BROOKLYN	NY	33	13	36	12PL	220.39		3F	C	1	15N	CRIM SALE CONTRL SUBST-3RD	100	100	10/06/03	23	1New York	BK	637	30000
06/05/098 MACON ST	BKLYN	NY	63	33	72	36PL	140.20		3F	D	0	7N	BURGLARY-3RD	101	140	06/28/89	35	1New York	BK	98	90000
02/15/9934 STERLING PL	BROOKLYN	NY	542	182	54	18PL	160.15		1F	C	1	5V	ROBBERY-1ST	100	100	07/30/03	30	1New York	BK	934	45000
04/20/0672 RIVERDALE AV	BROOKLYN	NY	33	13	36	12PL	160		0	Y	1	0	YO Adm Law Detail Not Known	100	100	11/06/03	0	1New York	BK	672	30000
07/03/0296B LEXINGTON AVE	BROOKLYN	NY	23	0	24	0PL	265.02		3F	D	0	20N	CRIM POSS WEAP-3RD DEG	100	100	11/13/03	23	1New York	BK	296	30000
12/08/02026 OCEAN AVE	BROOKLYN	NY	23	0	24	0PL	265.02		3F	D	0	20N	CRIM POSS WEAP-3RD DEG	100	100	04/17/03	23	1New York	BK	2026	30000
01/28/02057 80TH ST	BROOKLYN	NY	542	182	54	18PL	460.20		0F	B	0	43N	ENTERPRISE CORRUPTIO N	100	100	10/28/03	23	1New York	BK	2057	45000
04/29/0422 BLAKE AVE	BROOKLYN	NY	49	33	900	36PL	220.41		2F	A	0	15N	CRIM SALE CONTRL SUBST-2ND	100	100	11/04/03	23	1New York	BK	422	90000
05/02/0215 CLARKSON AVE	BROOKLYN	NY	63	23	72	24PL	220.16		3F	B	0	19N	CRIM POSS CONTR SUBST-3RD	100	100	07/18/03	0	1New York	BK	215	60000
08/22/0289 S 1 ST	BKLYN	NY	33	13	36	12PL	220.39		3F	B	0	15N	CRIM SALE CONTRL SUBST-3RD	100	100	08/01/03	0	1New York	BK	289	30000
03/06/0273 GATES AVE	BROOKLYN	NY	93	33	108	36PL	220.16		3F	B	0	19N	CRIM POSS CONTR SUBST-3RD	100	100	03/31/03	23	1New York	BK	373	90000
05/20/01043 MYRTLE AV	BROOKLYN	NY	63	23	72	24PL	220.31		5F	D	0	15N	CRIM SALE CONTRL SUBST-5TH	100	100	08/15/03	23	1New York	BK	1043	60000
11/26/0238 FORBELL STRE	BROOKLYN	NY	63	33	72	36PL	140.20		3F	D	0	7N	BURGLARY-3RD	100	140	12/10/98	0	1New York	BK	338	90000
04/24/0347 DEWITT AVE	BROOKLYN	NY	63	23	72	24PL	220.06		5F	D	0	19N	CRIM POSS CONTR SUBST-5TH	100	100	09/08/03	37	1New York	BK	347	60000
11/27/0282 SUNNYSIDE AV	BROOKLYN	NY	23	0	24	0PL	120.05		2F	D	0	6V	ASSAULT -2ND	100	100	10/30/03	23	1New York	BK	182	30000
07/13/091 EAGEL ST	BROOKLYN	NY	153	53	180	60PL	220.39		3F	B	0	15N	CRIM SALE CONTRL SUBST-3RD	100	100	01/16/03	0	1New York	BK	91	150000
04/19/9202 BAY 28TH ST	BKLYN	NY	93	542	108	54PL	140.25		2F	C	0	7V	BURGLARY-2ND	101	140	07/05/94	23	1New York	BK	202	135000
05/28/02157 WEST 5TH ST	BROOKLYN	NY	49	43	900	48PL	220.18		2F	A	0	19N	CRIM POSS CONTRL SUBST-2ND	100	100	12/01/03	0	1New York	BK	2157	120000
03/20/71570 E NEW YORK AVE	BROOKLYN	NY	49	153	900	180PL	125.25		0F	A	0	1L	MURDER	100	140	03/20/73	0	1New York	BK	1570	450000
11/25/064 BRIGHTON 1 P	BROOKLYN	NY	63	23	72	24PL	220.39		3F	B	0	15N	CRIM SALE CONTRL SUBST-3RD	100	100	01/24/03	23	1New York	BK	64	60000
01/30/0315 MARCY AVENUE	BROOKLYN	NY	43	0	48	0PL	120.10		1F	C	1	6V	ASSAULT 1ST DEGREE	100	100	05/05/03	23	1New York	BK	115	60000
03/17/0665 JEFFERSON AVE	BROOKLYN	NY	73	422	84	42PL	140.20		3F	D	0	7N	BURGLARY-3RD	100	141	12/29/98	62	1New York	BK	665	105000
02/03/991040 55ST	BKLYN	NY	302	0	30	0PL	160.10		2F	D	1	5V	ROBBERY-2ND	100	100	06/02/03	23	1New York	BK	1040	37500
08/26/02980 WEST 28 STRE	BROOKLYN	NY	43	162	48	16PL	220.41		2F	A	0	15N	CRIM SALE CONTRL SUBST-2ND	100	100	09/23/03	23	1New York	BK	2980	40000
04/30/0180 LENOX RD	BROOKLYN	NY	422	0	42	0PL	160.10		2F	C	0	5V	ROBBERY-2ND	100	100	08/26/03	0	1New York	BK	180	52500
12/17/01259 LORING AVENU	BROOKLYN	NY	43	23	48	24PL	220.06		5F	D	0	19N	CRIM POSS CONTR SUBST-5TH	100	100	06/03/03	0	1New York	BK	1259	60000
03/21/030 BERGEN ST	BKLYN	NY	33	182	36	18PL	155.30		4F	E	0	8N	GRAND LARCENY-4TH	101	101	02/25/03	0	1New York	BK	30	45000
06/07/018 DODWORTH STR	BROOKLYN	NY	422	0	42	0PL	160.15		1F	C	1	5V	ROBBERY-1ST	100	100	03/26/03	23	1New York	BK	18	52500
07/15/01237 DEKALB AVE	BROOKLYN	NY	73	0	84	0PL	160.15		1F	B	0	5V	ROBBERY-1ST	100	100	07/02/03	23	1New York	BK	1237	105000
04/23/029 AVENU E W	BROOKLYN	NY	33	33	36	36PL	220.34		4F	C	0	15N	CRIM SALE CONTRL SUBST-4TH	100	100	07/15/03	62	1New York	BK	29	90000
01/02/081 CORNE LIA STR	BROOKLYN	NY	33	13	36	12PL	220.39		3F	B	0	15N	CRIM SALE CONTRL SUBST-3RD	100	140	12/03/01	0	1New York	BK	81	30000
11/22/03160 FULTON STREE	BROOKLYN	NY	422	0	42	0PL	160.15		1F	C	1	5V	ROBBERY-1ST	100	100	05/28/03	0	1New York	BK	3160	52500
10/27/0198 HIMROD STREET	BROOKLYN	NY	33	182	36	18PL	140.20		3F	E	1	7N	BURGLARY-3RD	100	141	12/01/00	40	1New York	BK	198	45000
05/04/0112 MANHATTAN AV	BROOKLYN	NY	33	182	36	18PL	220.06		5F	E	1	19N	CRIM POSS CONTR SUBST-5TH	101	140	09/19/97	0	1New York	BK	112	45000
04/08/0160 LOTT AVENUE	BROOKLYN	NY	33	302	36	30PL	215.51	BVI	1F	E	0	37N	CRIM CONTEMPT-1ST:PHY MEN ANCE	100	130	12/07/99	0	1New York	BK	160	75000
08/28/02708 BED FORD AVEN	BROOKLYN	NY	33	182	36	18PL	130.65		1F	E	1	22N	SEXUAL ABUSE-1ST	100	100	10/01/03	30	1New York	BK	2708	45000
11/05/068 CUMBERLAND W	BROOKLYN	NY	33	182	36	18PL	155.30		4F	E	0	8N	GRAND LARCENY-4TH	100	100	12/29/03	23	1New York	BK	68	45000
02/21/01514 CARROLL ST	BROOKLYN	NY	43	162	48	16PL	160		0	Y	0	0	YO Adm Law Detail Not Known	100	100	02/21/03	23	1New York	BK	1514	40000
01/08/0370 BUSHWICK	BROOKLYN	NY	422	0	42	0PL	130.80	01B	2F	D	0	22V	COURSE SEX CONDUCT-2: CHILD<13	100	100	01/08/03	23	1New York	BK	370	52500
08/22/0109 MARCUS GARVE	BROOKLYN	NY	422	0	42	0PL	265.03		2F	C	0	20V	CRIM POSS WEAPON- 2ND DEGREE	100	100	08/29/03	23	1New York	BK	109	52500
12/11/033 WESTM INISTER	BROOKLYN	NY	103	422	120	42PL	220.39		3F	B	0	15N	CRIM SALE CONTRL SUBST-3RD	100	100	12/03/03	0	1New York	BK	33	105000
12/18/0919 MYRT LE AVENU	BROOKLYN	NY	53	302	60	30PL	220.31		5F	D	0	15N	CRIM SALE CONTRL SUBST-5TH	101	140	01/12/96	0	1New York	BK	919	75000
03/09/0157 GRAHAM AV	BKLYN	NY	43	23	48	24PL	220.31		5F	D	0	15N	CRIM SALE CONTRL SUBST-5TH	100	100	10/01/03	0	1New York	BK	157	60000
05/31/0351 5T H AVE	BKLYN	NY	93	542	108	54PL	220.16		3F	B	0	19N	CRIM POSS CONTR SUBST-3RD	100	100	01/16/03	23	1New York	BK	351	135000
10/15/0956 PARK PL	BKLYN	NY	43	6211	48	20PL	220.39		3F	B	0	15N	CRIM SALE CONTRL SUBST-3RD	100	100	03/19/03	0	1New York	BK	956	50000
06/06/0119 DEKA LB AVE	BKLYN	NY	33	9341	36	31PL	140.25		2F	D	1	7V	BURGLARY-2ND	101	101	05/08/03	0	1New York	BK	119	77500
07/06/01414 BERGEN STREE	BROOKLYN	NY	53	0	60	0PL	265.03		2F	C	0	20V	CRIM POSS WEAPON- 2ND DEGREE	100	100	11/19/03	23	1New York	BK	1414	75000
10/25/0429 DUMONT AVE 8D	BROOKLYN	NY	422	0	42	0PL	160.10		2F	C	0	5V	ROBBERY-2ND	100	100	03/31/03	23	1New York	BK	429	52500
10/15/0896 PARK AVE	BROOKLYN	NY	33	182	36	18PL	170.25		2F	E	1	24N	POSSESS FORGED INSTRUMENT -2ND	100	100	03/14/03	23	1New York	BK	896	45000
06/17/9444 RA LPH AVE	BROOKLYN	NY	93	542	108	54PL	220.39		3F	B	0	15N	CRIM SALE CONTRL SUBST-3RD	100	100	10/14/03	0	1New York	BK	444	135000
05/10/0494 MILF ORD STREET	BROOKLYN	NY	33	182	36	18PL	220.06		5F	E	1	19N	CRIM POSS CONTR SUBST-5TH	100	100	12/17/03	30	1New York	BK	494	45000
09/24/9101 WOODRUFF AVE	BKLYN	NY	63	33	72	36PL	165.50		3F	D	0	28N	CPSP-3RD-VALUE OF PROP > $30[illegible]	100	141	01/29/99	0	1New York	BK	101	90000
01/12/01283 PAC IFIC ST	BKLYN	NY	103	53	120	60PL	160.15		1F	C	1	5V	ROBBERY-1ST	101	140	07/06/93	23	1New York	BK	1283	150000
09/25/0560 PROS PECT PLA	BROOKLYN	NY	33	0	36	0PL	140.25		2F	D	1	7V	BURGLARY-2ND	100	100	04/11/03	23	1New York	BK	560	45000
01/22/0170 WYLCOFF AVE	BROKLYN	NY	43	162	48	16PL	220		0	Y	0	0	YO Adm Law Detail Not Known	100	100	09/15/03	23	1New York	BK	170	40000
03/27/034 ROCKAWAY AVE	BROOKLYN	NY	33	13	36	12PL	155.30		4F	E	0	8N	GRAND LARCENY-4TH	100	100	01/07/03	23	1New York	BK	34	30000
10/22/92676 LINDEN BLVD	BROOKLYN	NY	49	103	900	120PL	160.10		2F	C	0	5V	ROBBERY-2ND	101	140	04/08/81	23	1New York	BK	2676	300000
07/10/0270 JAMAICA AVE	BROOKLYN	NY	63	33	72	36PL	220.39		3F	C	1	15N	CRIM SALE CONTRL SUBST-3RD	101	140	11/20/95	0	1New York	BK	270	90000
08/28/0301 HINS DALE STR	BROOKLYN	NY	33	182	36	18PL	140.20		3F	E	1	7N	BURGLARY-3RD	100	100	02/27/03	23	1New York	BK	301	45000
03/06/0121 NOSTRAND AVE	BROOKLYN	NY	43	23	48	24PL	165.45		4F	E	0	28N	CRIM POSSESSION STOLN PROP-4TH	101	140	09/01/99	23	1New York	BK	121	60000
01/07/01258 ALBANY AVENU	BROOKLYN	NY	83	43	96	48PL	220.34		4F	C	0	15N	CRIM SALE CONTRL SUBST-4TH	101	101	10/24/03	0	1New York	BK	1258	120000
11/26/0682 MACDONOUGH S	BROOKLYN	NY	83	0	96	0PL	160.15		1F	B	0	5V	ROBBERY-1ST	100	100	07/08/03	23	1New York	BK	682	120000
02/11/0169-06 FOCH BOULEVA	BROOKLYN	NY	73	422	84	42PL	140.20		3F	D	0	7N	BURGLARY-3RD	100	140	03/06/97	0	1New York	BK	169-06	105000
08/01/01480 HERKIMER STR	BROOKLYN	NY	422	0	42	0PL	130.50		1F	C	1	22V	SODOMY-1ST	100	100	10/29/03	0	1New York	BK	1480	52500
06/06/0150 POWERS ST	BROOKLYN	NY	422	0	42	0PL	130.50		1F	C	1	22V	SODOMY-1ST	100	100	12/12/03	23	1New York	BK	150	52500
08/26/099 PILLING STRE	BROOKLYN	NY	43	0	48	0PL	160.10		2F	D	1	5V	ROBBERY-2ND	100	100	12/17/03	0	1New York	BK	99	60000
10/14/0730 EUCLID AVENU	BROOKLYN	NY	153	53	180	60PL	105.15		2F	B	0	43N	CONSPIRACY -2ND	100	100	08/12/03	23	1New York	BK	730	150000
05/18/0258 JEFFERSON AV	BROOKLYN	NY	43	0	48	0PL	120.05		2F	D	0	6V	ASSAULT -2ND	100	100	03/04/03	23	1New York	BK	258	60000
06/02/0610 SHEFFIELD AV	BROOKLYN	NY	23	0	24	0PL	140.25		2F	D	1	7V	BURGLARY-2ND	100	100	09/15/03	0	1New York	BK	610	30000
02/12/02805 W20TH ST	BROOKLYN	NY	53	0	60	0PL	160.15		1F	B	0	5V	ROBBERY-1ST	100	100	01/30/03	0	1New York	BK	2805	75000
02/18/0901 JEFFERSON AV	BKLYN	NY	302	0	30	0PL	160.10		2F	D	1	5V	ROBBERY-2ND	100	100	01/17/03	0	1New York	BK	901	37500
10/13/0317 4 AVENUE	BROOKLYN	NY	33	13	36	12PL	220.18		2F	A	0	19N	CRIM POSS CONTRL SUBST-2ND	100	100	07/22/03	23	1New York	BK	317	30000
07/29/0328 FLATBUSH AVE W	BROOKLYN	NY	33	13	36	12PL	160		0	Y	0	0	YO Adm Law Detail Not Known	100	100	07/29/03	23	1New York	BK	328	30000
02/12/02805 W 20TH ST	BROOKLYN	NY	53	0	60	0PL	160.15		1F	B	0	5V	ROBBERY-1ST	100	100	01/24/03	0	1New York	BK	2805	75000
08/26/0115 AVE.I	BROOKLYN	NY	49	203	900	240PL	125.25		2F	A	0	1L	MURDER-2ND DEG	100	100	01/08/03	23	1New York	BK	115	600000
01/26/0635 STERLING PL	BROOKLYN	NY	53	0	60	0PL	120.10		1F	B	0	6V	ASSAULT 1ST DEGREE	100	100	07/17/03	23	1New York	BK	635	75000
01/08/0311 SOUTH 2 STRE	BROOKLYN	NY	103	1022	120	102PL	125.25		2F	B	1	1V	MURDER-2ND DEG	100	100	07/15/03	23	1New York	BK	111	255000
10/27/0433 LAFAYETTE AVE	BROOKLYN	NY	33	182	36	18PL	155.30		4F	E	0	8N	GRAND LARCENY-4TH	100	100	07/31/03	23	1New York	BK	433	45000
01/14/0354 ELTON STREET	BROOKLYN	NY	33	13	36	12PL	265		0	Y	1	0	YO Adm Law Detail Not Known	100	100	09/25/03	23	1New York	BK	354	30000
04/25/0512 OCEANVIEW AV	BROOKLYN	NY	49	43	900	48PL	220.18		2F	A	0	19N	CRIM POSS CONTRL SUBST-2ND	100	100	03/06/03	23	1New York	BK	512	120000
06/30/0130 3RD AVE	BROOKLYN	NY	33	13	36	12PL	160		0	Y	0	0	YO Adm Law Detail Not Known	100	100	06/30/03	23	1New York	BK	130	30000
09/07/0554 FLUSHIGN AV	BKLYN	NY	93	33	108	36PL	220.16		3F	B	0	19N	CRIM POSS CONTR SUBST-3RD	100	100	04/28/03	23	1New York	BK	554	90000
12/21/013 SCAFFER ST	BKLYN	NY	422	0	42	0PL	160.10		2F	C	0	5V	ROBBERY-2ND	100	100	01/31/03	0	1New York	BK	13	52500
07/13/0108 KINGSBOROUGH	BROOKLYN	NY	63	0	72	0PL	125.25		2F	B	1	1V	MURDER-2ND DEG	100	100	05/28/03	0	1New York	BK	108	90000

Excerpt from a database of New York City prisoners by home address with expenditures added [data has been scrambled].

MONEY MAPS

Measured in dollars, the criminal justice network has frequently become the most important public institution in high-resettlement neighborhoods. The stakes and impacts of this unacknowledged investment become clearer when we make the incarceration maps slightly more complex by adding information about the actual costs of imprisonment. How much money does it cost to keep people in prison? The figures are available, and when they are correlated with the addresses of the people on whom the money is being spent, a remarkable pattern emerges.

We call them "million-dollar blocks": single blocks in inner-city neighborhoods across the country for which upward of a million dollars is allocated each year to imprison its residents.

The maps now suggest a link between those places and the dollars spent (elsewhere) on their residents. They ask us to weigh the opportunity costs—for each city block, neighborhood, or wider community—of committing those funds to recycle people through jail and prison, back home, and then (for more than a third of them) back inside again. This pattern is visible in all too many major American cities: New Haven, New Orleans, New York City, Phoenix, and Wichita.

Money spent on criminal justice is money not spent on other civic institutions, especially in these communities. Guided by the maps of million-dollar blocks, urban planners, designers, and policy makers can identify those areas in our cities where—without acknowledging it—we have allowed the criminal justice system to replace and displace a whole host of other public institutions and civic infrastructure. Those neglected sectors are the very ones we have already identified as the collateral damage of the incarceration explosion: education, family, housing, health, civic involvement. Now the investment pattern and spending priorities that feed this condition become dramatically evident.

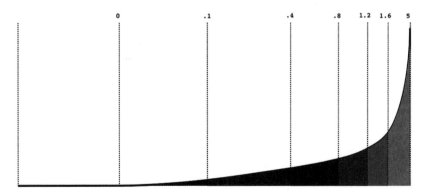

Prison expenditures expressed in millions of dollars: The resulting histogram displays what statisticians call a Power Law distribution, in which the largest share of the total expenditure is represented by a very small share of census blocks.

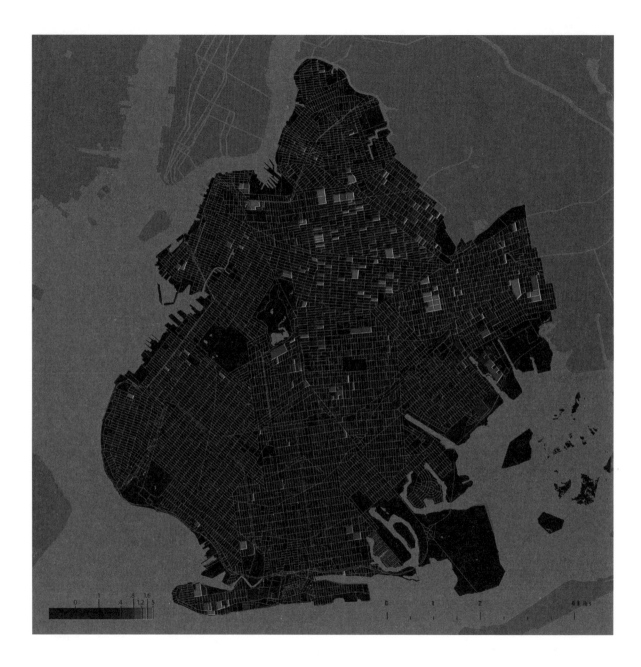

Prison expenditures by census block in Brooklyn,
New York, 2003.

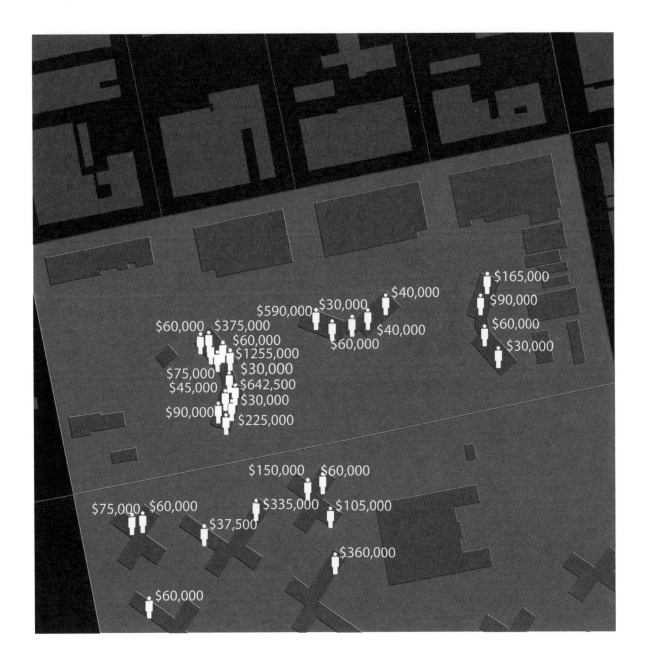

Thirty-one men, 4.4 million dollars, four blocks of Brownsville, Brooklyn, New York, 2003.

CRIMINAL JUSTICE AS INFRASTRUCTURE

No matter how physically removed they are from the neighborhoods of the people they hold, the urban exostructure of prisons and jails remain firmly rooted as institutions of the city, as everyday parts of life for people, affecting their homes, social networks, and movements.

An analysis of any million-dollar block will demonstrate how the overlapping resources of these networks conflate individuals and infrastructure, the local and the global, the close and the far, the piece and the system. Doing anything here— attempting to restructure the way the criminal justice system works—means working with contingent, dynamic, and overlapping systems and collaborations between multiple agencies, tools, and techniques.

What does it mean to design policy, to design multiple policies, around a single place?

The maps are both a picture and a design strategy. The picture is an aggregate situation. The design strategy is "start from the block and build," incrementally, new networks that might inform this crippled urban infrastructure.

In this way, these maps depart radically from the maps and statistical analyses that fueled mid-twentieth-century efficient city, urban renewal, and policing projects. The map is not a top-down view. And neither is it a bottom-up account. It is both.

Identify an area. Zoom in and examine the specific conditions. Zoom out and then consider both scales at the same time. The resulting image is no longer hard data. It is a soft map that is infinitely scalable, absolutely contingent, open to vision and hence revision.

Acknowledgments

This book took too long to finish. The usual things got in the way: new projects, life, work duties, distractions, all the right things. It would be nice to say that the length of time it required means that everything I've been doing is included here. But, in fact, it's a selection, a collection of a certain kind of work with maps, data, and spatial imagery. As the geo-web rapidly expands into new directions and media—from participatory mapping, to locative arts, to crowd sourced maps, to mobile platforms—this is a short pre-history to the current explosion.

Embedded as this work has been in new technologies, higher resolutions, and current events, its form constantly shifted. As soon as any project was finished, it was ready for reframing. The book is a collection of digital artifacts, transferred from floppy disks to zip disks, Jazz drives to CDs, and external drives to the cloud. It has survived many operating systems and software updates.

Thanks are due to many institutions and people. First, to Michel Feher for waiting for the book, and to the Zone team: Julie Fry for the careful attention that she has paid to the images and to the design of the book, Meighan Gale for her attention to absolutely everything, and Bud Bynack for his generous editing and sense of humor.

I owe an enormous debt of gratitude to Kyong Park and Shirin Neshat at the Storefront for Art and Architecture, who took the risk of giving me my first solo show in New York, when all I had to show them were a few slides of a GPS receiver that I did not yet have in hand, and some military images that had been drawn with this device. Xavier Costa understood, after seeing the show at the Storefront, that I needed a bigger palate for the GPS drawings and invited me to remake the exhibit in a gallery of MACBA in Barcelona for its inaugural show. In many ways these two exhibits set the direction of my form of spatial practice.

Many other curators and organization directors have since supported the projects outlined in this book: Paola Antonelli, Ursula Frohne, Rosalie Genevro, Carles Guerra, Jürgen Harten, K. Michael Hays, Thomas Y. Levin, Patricia McDonald, Jennifer Riddell, Diane Shamash, Harald Szeemann, and Rob Silberman, among others. Their support in many cases enabled the purchase, production, and display of the digital imagery. Museums and not-for-profit galleries have been important and generous institutions in the development of my project. Those institutions have long lives, but it is always specific people who have worked hard with me to make the projects come about. I am grateful to them.

I have also been lucky to receive a number of grants. The Graham Foundation funded my very first experiments with digital mapping, and at the very end of this project helped enable the production of this book. The Princeton University Humanities Council, the Open Society Institute, the JEHT Foundation, and the Rockefeller Foundation have all generously supported this work. A United States Artists Rockefeller Fellowship in 2009 catalyzed the completion of this volume.

Thanks to Gavin Browning, who sorted through the twenty-five or so introductions I had in a folder called "BookIntro" on my computer in 2010. He was enormously helpful in thinking about what should be included and excluded in those many versions, and in structuring the organization of the book as a whole.

The kind of work I do is never done alone, and I have had a series of wonderful collaborators, students, and production crews who have worked with me on all sorts of installations and exhibitions. They are named in the Exhibitions, Installations, Publications section below. Although you won't find her name in that section, Jenna Miller provided crucial assistance with the production of this book.

Thanks as well to my colleagues at Columbia and Princeton Universities, where most of this work was completed, for providing such productive academic environments. I owe intellectual debts to many colleagues and students who helped me to think about my work on trains, on subways, in cars, at lunch, at dinner, in the hallways, in academic reviews, and beyond: Christine Boyer, Beatriz Colomina, Elizabeth Diller, Hal Foster, Mario Gandelsonas, Leslie Gill, Laurie Hawkinson, Janette Kim, Andrea Khan, Thomas Levin, Ralph Lerner, Peter Marcuse, Reinhold Martin, Helen Nissenbaum, Kate Orff, Antoine Picon, David Reinfurt, Lindy Roy, Felicity Scott, Kendall Thomas, Mark Wasiuta, and Mabel Wilson. Since 2004 most of my work has been done at the Spatial Information Design Lab; I am grateful to my Co-Director and collaborator Sarah Williams and to Eric Cadora for introducing me to Million-Dollar Blocks. Thanks especially to Mark Wigley for providing such an amazing atmosphere of collegiality and support at Columbia.

Thank you as well to all those who have helped me think about my work, written about it, or published it: Janet Abrams, Emily Apter, Geoffrey Batchen, Kadambari

Baxi, Noah Chasin, Denis Cosgrove, Teddy Cruz, Olivia Custer, Rosalyn Deutsche, Keller Easterling, Paul Elliman, Mark Hansen, Peter Hall, Sarah Herda, Catherine Ingraham, John Ippolito, Natalie Jeremijenko, Branden Joseph, Miwon Kwon, Sylvia Lavin, Tom McDonough, Ricardo Scofidio, Vivian Selbo, Jesse Shapins, Susan Tucker, Lynne Tillman, Anthony Vidler, and Eyal Weizman.

Finally, my unbounded thanks goes to Tom Keenan who is the fiercest constructive critic of my work. Most people who know him know his generosity, kindness, and intellectual curiosity. He has been extremely forgiving, patient, and always willing to read my work, in and amongst his multiple obligations. To Claire, there will never be enough words for her in this format. I'm looking forward to more weekends together once this book goes to press. To my parents, who have always supported and even admired my work, although they claim they don't always understand it. To my sister, thanks for being there, always, even though you're on the other side of the world.

YOU ARE HERE: MUSEU. Computer-based installation
utilizing Global Positioning System mapping technology.
Inaugural installation, Department of Architecture
and Design, Museu d'Art Contemporani de Barcelona,
November 1995 – February 1996.

Exhibitions, Installations, Publications

1 YOU ARE HERE

YOU ARE HERE: INFORMATION DRIFT. Computer-based installation utilizing Global Positioning System mapping technology. Storefront for Art and Architecture, New York, NY, March–April 1994. Curators: Kyong Park and Shirin Neshat. Computer imaging and installation: Robert Brawner, Sergio Bregante, Kevin Kunkel, James Luhur, Amy Nowacki, Mark Sanchez, Angela Utschig. Technical consultants: Sue Fierro, Scott Patterson, Darrel Peterson, Ken Salsnan, Jeff Teitelbaum, and Alan White.

YOU ARE HERE: MUSEU. Computer-based installation utilizing Global Positioning System mapping technology. Inaugural installation, Department of Architecture and Design, Museu d'Art Contemporani de Barcelona, November 1995 – February 1996. Curator: Xavier Costa. Project Coordinator: Anna Guarro. Computer imaging assistance: Ron De Villa.

"You Are Here: Information Drift, 12 March to 16 April, Laura Kurgan," Storefront for Art and Architecture, March 1994; "You Are Here: Information Drift," *Assemblage* 25 (December 1994); "You Are Here: Museu," in Laura Kurgan and Xavier Costa, ed., *You Are Here: Architecture and Information Flows* (Barcelona, MACBA, 1995).

2 KUWAIT: IMAGE MAPPING

"You Are Here: Kuwait," *Documents* 1/2 (Fall/Winter 1992).

3 CAPE TOWN, SOUTH AFRICA, 1968: SEARCH OR SURVEILLANCE?

CLOSE UP AT A DISTANCE. Installation using declassified high-resolution Corona satellite imagery of Cape Town, South Africa, 1968. Exhibited in "The Art of Detection: Surveillance in Society," List Visual Arts Center, Massachusetts Institute of Technology, Cambridge, October–December 1997. Curator: Jennifer Riddell. Re-exhibited in "Transatlantico," Centro Atlantico de Arte Moderne, Las Palmas de Gran Canaria (Canary Islands), April–June 1998. Curator: Octavio Zaya.

"Close up at a Distance," in Jennifer Riddell et al., *The Art of Detection: Surveillance in Society* (Cambridge, MA: MIT List Visual Arts Center, 1997); "Primer plano a distanca," in Octavio Zaya et al., *Transatlantico* (Cabldo de Gran Canaria: Centro Atlantico de Arte Moderne, 1998).

4 KOSOVO 1999: SPOT 083-264

SPOT 083-264, KOSOVO, JUNE 3, 1999. Installation using SPOT commercial satellite imagery to investigate a mass grave site in Kosovo during the NATO air campaign. Exhibited in "World Views: Maps and Art," Weisman Art Museum, University of Minnesota, Minneapolis, September 1999–January 2000. Curators: Rob Silberman in collaboration with Patricia McDonnell. Re-exhibited in "Kosov@," Pacific Northwest College of Art, Portland, Oregon, April 2000. Curator: Trebor Scholz. Re-exhibited in "Anxious Omniscience," Princeton University Art Gallery, Princeton, New Jersey, February–April 2002. Curator: Thomas Y. Levin.

"SPOT 083-264: Kosovo, June 3, 1999," in Robert Silberman with Patricia McDonnell, *World Views: Maps & Art* (Minneapolis and London: University of Minnesota Press, 1999); "SPOT 083-264: Kosovo, June 1999," *Archis* 2 (2000); "Inadvertent Memory," *Cabinet* 1 (Winter 2000/01).

5 NEW YORK, SEPTEMBER 11, 2001

NEW YORK, SEPTEMBER 11, 2001, FOUR DAYS LATER... Installation using high-resolution Ikonos satellite imagery of New York City, on September 15, 2001. Exhibited in "CTRL [SPACE]: Rhetorics of Surveillance from the Panopticon to Big Brother," Zentrum für Kunst und Medien, Karlsruhe, October 2001–February 2002. Curator: Thomas Y. Levin. Project Assistance: Janette Kim. Another version was installed in "911+1: The Perplexities of Security," Watson Institute for International Studies, Brown University, Providence, Rhode Island, September 2002. Curator: James Der Derian.

"New York, September 11, 2001, Four Days Later...," in Thomas Y. Levin, Ursula Frohne, and Peter Weibel, eds., *CTRL* [*SPACE*]: *Rhetorics of Surveillance from Bentham to Big Brother* (Karlsruhe: ZKM Center for Art and Media, and Cambridge, MA: The MIT Press, 2002)

6 AROUND GROUND ZERO

AROUND GROUND ZERO. Fold-out map of the area around the former site of the World Trade Center, intended to orient visitors and facilitate the work of remembrance; in two editions, December 2001 and March 2002. Project team: Janette Kim and Bethia Liu, with participation by Rivka Mazar and Donald Shillingburg. Photography: Margaret Morton. The following organizations supported printing of the fold out maps: American Institute of Architects (New York Chapter), Con Edison, the New York Community Trust, the New York Foundation of the Arts, the Open Society Institute, Princeton University, TBWA Chiat Day, the Van Alen Institute, and two anonymous donors.

"Around Ground Zero," New York: New York New Visions, December 2001, 2nd ed. March 2002; also distributed in *Grey Room* 07 (Spring 2002); "Around Ground Zero: interview with Alice Twemlow," *Trace* 1.4 (2003).

7 MONOCHROME LANDSCAPES

MONOCHROME LANDSCAPES. Installation using high-resolution Ikonos and Quick-Bird satellite imagery of Alaska, the Atlantic Ocean, Cameroon, and Iraq. Exhibited in "Architecture by Numbers," Whitney Altria, New York, March–May 2004. Curator: K. Michael Hays. Image processing: Bill Guthe and Erik Carver. Supported by a grant from the Committee on Research in the Humanities and Social Sciences, Princeton University. Re-exhibited "On The Wall: Aperture magazine '05–'06," Aperture Photography Gallery, New York, January–March 2007. Curator: Melissa Harris.

SHADES OF GREEN: Exhibition using high-resolution Ikonos and GeoEye satellite imagery of rain forest destruction in Brazil and in Indonesia. Exhibited in "Antiphotojournalism," La Virreina Centre de l'Imatge, Barcelona, July–October 2010 and at Foam, Amsterdam, April–June 2011. Curators: Carles Guerra and Thomas Keenan. Forestry advisers: François-Michel Le Tourneau, French National Center for Scientific Research (CNRS), unit for Research and Scientific Information on the Americas and Susan Minnemeier, World Resources Institute. Image processing: Maria Cavaller and Marc Roig.

SHADES OF GREEN. Exhibition using high-resolution Ikonos and GeoEye satellite imagery of rain forest destruction in Brazil and in Indonesia. Exhibited in "Antiphotojournalism," La Virreina Centre de l'Imatge, Barcelona, July–October 2010 and at Foam, Amsterdam, April–June 2011.

"Monochrome Landscapes," in Janet Abrams and Peter Hall, eds., *Else/Where: Mapping* (Minneapolis: University of Minnesota Design Initiative, 2006). "Monochrome Landscapes," in Janet Abrams and Peter Hall, eds., *Else/Where: Mapping* (Minneapolis: University of Minnesota Design Initiative, 2006); "Laura Kurgan," in *Antiphotojournalism 1.4–8.6.2011* (Amsterdam: Foam, 2011).

8 GLOBAL CLOCK

SHORT CIRCUIT. Computer-based installation utilizing real time financial data feeds to explore the time and space of foreign exchange (digital yen, Euros, and dollars, as traded in global markets). Exhibited in "The Fifth Element," Kunsthalle Düsseldorf, January – May 2000. Curator: Jurgen Harten. Javascript programming: David Frackman and Johnna Cressica Brazier. Re-exhibited as Global Clock No. 1, 1999/2000, in "Empire/State, Artists Engaging Globalization," Whitney Museum of American Art, Independent Study Program Exhibition, Art Gallery of the Graduate Center of the City University of New York, May – July 2002.

GLOBAL CLOCK NO. 2, 2002. Clock driven by real time data feeds from financial and stock exchange markets. Exhibited in "Money and Value," in Biel/Bienne,

May–October 2002, part of EXPO.02, Swiss National Exposition. Curator: Harald Szeemann. Programming and visualization: Erik Carver and Sean Dockray.

"Short Circuit," in Jürgen Harten, ed., *Das Fünfte Element: Geld oder Kunst* (Düsseldorf: Kunsthalle Düsseldorf, and Köln: DuMont Buchverhandlung, 2000); "Global Clock No. 2," in Harald Szeemann et al., *Money and Value—the Last Taboo* (Zürich: Swiss National Exhibition and Edition Oehrli, 2002); part of the introduction to this chapter was published in "Trying Not to Avoid Propositions, Altogether," *Assemblage* 41 (April 2000).

9 MILLION-DOLLAR BLOCKS

ARCHITECTURE AND JUSTICE. Data maps of "million-dollar blocks," demonstrating emerging patterns in cities across the United States. Architecture League of New York, September–October 2006. A project of the Spatial Information Design Lab, Graduate School of Architecture, Planning and Preservation, Columbia University. Project Directors: Laura Kurgan and Eric Cadora. Research associates: Sarah Williams and David Reinfurt. Research assistant: Leah Meisterlin. Re-exhibited in "Design and the Elastic Mind," Museum of Modern Art, New York, February 2008. Curator: Paola Antonelli. Re-exhibited in "JUST SPACE(S)," Los Angeles Contemporary Exhibitions, October 2007. Curator: Nicholas Brown.

Spatial Information Design Lab, *Architecture and Justice*, Columbia University Graduate School of Architecture, Planning, and Preservation, 2008; and in Paola Antonelli, ed., *Design and the Elastic Mind* (New York: Museum of Modern Art, 2006).

Notes

INTRODUCTION

1. Denis Cosgrove, *Apollo's Eye: A Cartographic Genealogy of the Earth in the Western Imagina-tion* (Baltimore: Johns Hopkins University Press, 2001), p. 257.

2. Quoted by Cosgrove, *Apollo's Eye*, p. 258.

3. Martin Heidegger, "'Only a God Can Save Us': The *Spiegel* Interview (1966)," trans. William J. Richardson, S.J., in Thomas Sheehan (ed.), *Heidegger: The Man and the Thinker* (Chicago: Precedent Publishing, 1981), p. 56.

4. Cosgrove, *Apollo's Eye*, p. 257.

5. NASA Earth Observatory, "History of the Blue Marble," undated, http://earthobservatory. nasa.gov/Features/BlueMarble/BlueMarble_history.php.

6. NASA Goddard Space Flight Center, "VIIRS Eastern Hemisphere Image — Behind the Scenes," February 2, 2012, http://www.nasa.gov/topics/earth/features/viirs-globe-east.html.

7. *The Blue Marble* has been an object of fascination for many. Al Gore proposed in 1998 to position a satellite at one of the Earth's Lagrangian points, where the satellite would always capture a sun-lit view of the Earth. Gore wanted to make this image continuously available in real time on the Internet. "As White House aides tell it, Mr. Gore's idea came to him almost in a dream. He awoke at 3 a.m. one day in February 1998, and thought about a global view of Earth, they say, perhaps inspired by the famous 'Blue Marble' photograph of the planet taken in December 1972 by the returning crew of Apollo 17, the last moon mission. A copy of the picture hangs in Mr. Gore's office." The fantastical idea was never realized. See Warren E. Leary, "Politics Keeps a Satellite Earthbound," *New York Times*, June 1, 1999, p. F1.

8. See Jeff Richardson, "Blue Marble," *iPhone J.D.*, March 10, 2010, http://www.iphonejd.com/ iphone_jd/2010/03/blue-marble.html, and NASA Earth Observatory, "History of the Blue Marble."

9. In this sense, I take some distance, so to speak, from the work of Lisa Parks, whose *Cultures in Orbit: Satellites and the Televisual* (Durham: Duke University Press, 2005) is an excellent study of the culture of satellite images in politics, the news, and our imagination. Her inter-pretations, however, stop at the ways in which these images are presented to us, which is to say that she reads only what has already been interpreted. The same can be said for other

influential theorists of satellite space, such as Paul Virilio, and sky-friendly geographers such as Denis Cosgrove and J.B. Harley. Although I have learned much from all them, the driving force behind the work in this book is somewhat different: not simply to talk about these images, but to work with them, do things with them, help create them, actively shape, produce, and modify them.

10. Peter Galison, *Einstein's Clocks, Poincaré's Maps* (New York: W. W. Norton, 2003), p. 285.

11. *Ibid.*, p. 287.

12. *Ibid.*, pp. 288–89.

13. *Ibid.*, pp. 292–93.

14. Rosalyn Deutsche, "Boys Town," *Environments and Planning D, Society and Space* 9.1 (March 1991), p. 21.

15. Philip Morrison and Phyllis Morrison and the Office of Charles and Ray Eames, *Powers of Ten: About the Relative Size of Things in the Universe* (New York: Scientific American Library, 1982).

16. *Ibid.*, pp. iv–v.

17. *Ibid.*, p. 145.

18. See "Satellite Images Offer Detailed Views from Space," transcript, Neal Conan interview with John Pike of GlobalSecurity.org, *Talk of the Nation*, National Public Radio, July 10, 2007, http://www.npr.org/templates/transcript/transcript.php?storyId=11850958. Pike says: "Google is not independently acquiring the imagery. They're getting it from these commercial companies—two in the United States, one in Israel, one in France. The American companies are operating under government license. And so there are some restrictions on what they can acquire imagery of and how they can release it. So the Defense Department has already signed off on the public release of this imagery before Google is able to acquire it."

19. Barbara Crossette, "U.S. Seeks to Prove Mass Killings," *New York Times*, August 11, 1995, p. A3. The withholding of the images, especially the ones acquired by satellites, from the general public was challenged in a Freedom of Information Act lawsuit by Students Against Genocide. See the remarkable decision by the Court of Appeals for the District of Columbia Circuit in *Students Against Genocide v. Department of State*: http://www.cadc.uscourts.gov/internet/opinions.nsf/A26612A99D23C92B85256F7A0064436F/$file/99-5316a.txt.

20. See, on declassification, Kevin C. Ruffner, "CORONA and the Intelligence Community: Declassification's Great Leap Forward," April 14, 2007, Central Intelligence Agency, https://www.cia.gov/library/center-for-the-study-of-intelligence/csi-publications/csi-studies/studies/96unclass/corona.htm. Of course, although as John Pike says, "the satellite genie seems now irreversibly out of the bottle," the same cannot be said for American reconnaissance and surveillance programs in general, as the policies of the Bush Administration were to demonstrate quite dramatically. Pike, in "Satellite Images Offer Detailed Views from Space," *Talk of the Nation*, National Public Radio, July 10, 2007.

21. "The World: Campaign Poster; Grozny, Dec. 16, 1999 / Grozny, March 16, 2000," *New York Times*, March 26, 2000, section 4, p. 1, http://www.nytimes.com/2000/03/26/weekinreview/the-world-campaign-poster-grozny-dec-16-1999-grozny-march-16-2000.html.

22. Lara J. Nettelfield, "Terror in Chechnya: Russia and the Tragedy of Civilians in War" (review), *Human Rights Quarterly* 33. 3 (August 2011), p. 886.

23. Robert Mackey, "Satellite Image Shows North Korean Rocket Launch," *New York Times*, *The Lede* blog, April 8, 2009, http://thelede.blogs.nytimes.com/2009/04/08/satellite-image-shows-north-korean-rocket-launch.

24. Ian Sample, "North Korean Rocket Launch Caught on Film," *Guardian*, April 8, 2009, p. 21, http://www.guardian.co.uk/world/2009/apr/08/north-korea-rocket-launch-image.

25. Google Earth imagery is not entirely free: it is paid for by spatially motivated advertising. For a fuller argument on this, see Chris Perkins and Martin Dodge, "Satellite Imagery and the Spectacle of Secret Spaces," *Geoforum* 40 (2009), pp. 546–60, http://personalpages.manchester.ac.uk/staff/m.dodge/cv_files/spectacle_of_secret_spaces.pdf.

26. See the study from the RAND Corporation and the American Society of Photogrammetry and Remote Sensing, John Baker, Kevin N. O'Connell, and Ray A. Williamson (eds.), *Commercial Observation Satellites: At the Leading Edge of Global Transparency* (Santa Monica: Rand Publishing, 2001), p. vii.

27. William J. Broad, "Giant Leap for Private Industry: Spies in Space," *New York Times*, October 13, 1999, p. A14, http://www.nytimes.com/1999/10/13/us/giant-leap-for-private-industry-spies-in-space.html.

28. Quoted in William J. Broad, "Private Spy in Space to Rival Military's," *New York Times*, April 27, 1999, p. F1, http://www.nytimes.com/1999/04/27/science/private-spy-in-space-to-rival-military-s.html.

29. Ann M. Florini and Yahya A. Dehqanzada, "Secrets For Sale: How Commercial Satellite Imagery Will Change the World," Washington, D.C.: Carnegie Endowment for International Peace, 2000, http://www.carnegieendowment.org/publications/index.cfm?fa=view&id=160.

30. Secretary Colin L. Powell, "Remarks to the United Nations Security Council," New York City, February 5, 2003 (including slide show), http://www.globalsecurity.org/wmd/library/news/iraq/2003/iraq-030205-powell-un-17300pf.htm. The transcript is also available at "Threats and Responses. Powell's Address, Presenting 'Deeply Troubling' Evidence on Iraq," *New York Times*, February 6, 2003, p. A18, http://www.nytimes.com/2003/02/06/world/threats-responses-powell-s-address-presenting-deeply-troubling-evidence-iraq.html.

31. Open Source Center, "Iraqi Insurgency Group Utilizes 'Google Earth' for Attack Planning," July 19, 2006, http://www.fas.org/irp/dni/osc/osc071906.pdf; see also Peter Eisler, "Google Earth Helps Yet Worries Government, *USA Today*, November 7, 2008; and Open Source Center, "The Google Controversy—Two Years Later," July 30, 2008, http://www.fas.org/irp/dni/osc/google.pdf.

32. Satellite Sentinel Project: Monitoring the Crisis in Sudan, "Documenting the Crisis," http://satsentinel.org/documenting-the-crisis.

33. *Ibid*.

34. Lisa Parks, *Cultures in Orbit: Satellites and the Televisual* (Durham: Duke University Press, 2005), pp. 79, 77.

35. *Ibid*., p. 81. She offers what are in effect revisionist accounts of the massacre ("Islamic fundamentalists provoked the massacre as an act of martyrdom and then killed themselves," etc.), which she prophylactically qualifies under the headings of "each political interest puts its own spin on the event" and "the impossibility of knowing exactly what happened at Srebrenica in July 1995," but which nonetheless seem to suggest that she has her doubts about the standard accounts—accounts that, I should add, have been confirmed at the International Criminal Tribunal for the Former Yugoslavia in The Hague and accepted even by most critics of U.S. policy during and after the war.

36. *Ibid*., pp. 84, 90.

37. *Ibid*., pp. 90–91.

38. *Ibid.*, p. 101.

39. *Ibid.*, pp. 106–107.

40. *Ibid.*, p. 101.

41. Pike, in "Satellite Images Offer Detailed Views from Space," National Public Radio, July 10, 2007.

42. J. B. Harley, "Deconstructing the Map," in *The New Nature of Maps: Essays in the History of Cartography*, ed. Paul Laxton (Baltimore; Johns Hopkins University Press, 2001), pp. 151–52.

43. *Ibid.*, pp. 165, 166, 157. See also the discussion of the Peters projection versus the Mercator projection in Mark Monmonier, *Drawing the Line: Tales of Maps and Cartocontroversy* (New York: Henry Holt, 1995), pp. 9–25.

44. Svetlana Alpers, "The Mapping Impulse in Dutch Art," in *The Art of Describing: Dutch Art in the Seventeenth Century* (Chicago: University of Chicago Press, 1983), p. 138; Thomas Nagel, *The View from Nowhere* (New York: Oxford University Press, 1986).

45. *Ibid.*

46. Harley, "Deconstructing the Map," p. 166. The formalization and abstraction of the view from nowhere can allow us to ignore the politics underlying the development of technology, mathematics, and science that have enabled its production.

47. Yve-Alain Bois, "Metamorphosis of Axonometry," *Daidalos* 1.15 (September 1981), pp. 40–58.

48. *Ibid.*, pp. 56–57.

49. *Ibid.*, p. 57.

50. Cosgrove, *Apollo's Eye*, pp. 239 and 242. Paul Virilio, in *War and Cinema: The Logistics of Perception* (London: Verso, 1989), also explores the "logistics of military perception." Also essential in recounting this history is Beaumont Newhall, *Airborne Camera: The World from the Air and Outer Space* (New York: Hastings House, 1969).

51. In an essay titled "'The View from Nowhere'?: Spatial Politics and the Cultural Significance of High-Resolution Satellite Imagery," introducing an issue of *Geoforum*, Martin Dodge and Chris Perkins talk about the view from nowhere as an objective view. They say: "The papers in this collection question what can be seen in high-resolution satellite imagery and what this might mean. This work can perhaps be best introduced by considering the seductively objective view of the world that they claim to represent—what one might term, following Thomas Nagel (1986), the 'view from nowhere'. The key question is how this view is manufactured and mediated." All the papers in that collection address the point of view of high-resolution satellites. *Geoforum* 40.4 (2009), http://personalpages.manchester.ac.uk/staff/m.dodge/cv_files/view_from_nowhere_intro.pdf, p. 2.

52. Bruno Latour, "From Realpolitik to Dingpolitik, or How to Make Things Public," in Bruno Latour and Peter Weibel (eds.), *Making Things Public: Atmospheres of Democracy*, trans. Robert Bryce et al. (Cambridge, MA: The MIT Press, 2005), p. 26.

LEXICON

1. GPS.gov: Official U.S. Government Information about the Global Positioning System (GPS) and Related Topics, http://www.GPS.gov.

2. Garmin, "What is GPS?," http://www8.garmin.com/aboutGPS; United States Naval Observatory, "USNO NAVSTAR Global Positioning System," http://tycho.usno.navy.mil/gpsinfo.html.

3. GPS.gov, "GPS Accuracy," http://www.gps.gov/systems/gps/performance/accuracy.

4. For complete list of policy decisions, see National Executive Committee for Space-Based Positioning, Navigation, and Timing, "U.S. Space-Based PNT Policy," http://pnt.gov/policy.

5. The White House Office of Science and Technology Policy National Security Council, "Fact Sheet, U.S. Global Positioning System Policy," March 29, 1996, http://www.navcen.uscg.gov/pdf/1996_GPS_Policy.pdf.

6. National Executive Committee for Space-Based Positioning, Navigation, and Timing, "U.S. Space-Based Positioning, Navigation, and Timing Policy," December 15, 2004, http://www.pnt.gov/policy/2004-policy.shtml.

7. For a complete inventory of currently orbiting remote sensing satellites by country as of 2010, see Federation of American Scientists, Intelligence Resource Program, *The Remote Sensing Tutorial: What You Can Learn From Sensors on Spacecraft That Look Inward at the Earth and Outward at the Planets, the Galaxies and, Going Back in Time, The Cosmos*, http://www.fas.org/irp/imint/docs/rst/Front/overview2.html, and Socioeconomic Data and Applications Center, "An Annotated Guide to Earth Remote Sensing Data and Information Resources for Social Science Applications, Section 2.0: Selected Listings of Earth Remote Sensing Data and Information Resources, http://sedac.ciesin.columbia.edu/remote/app2.html.

8. Chris Perkins and Martin Dodge, "Satellite Imagery and the Spectacle of Secret Spaces," *Geoforum* 40.4 (2009), pp. 546–60, http://personalpages.manchester.ac.uk/staff/m.dodge/cv_files/spectacle_of_secret_spaces.pdf.

9. For the best analysis of these issues see Ann M. Florini and Yahya A. Dehquanzada, "No More Secrets?: Policy Implications of Commercial Remote Sensing Satellites," Carnegie Paper 1 (July 1999), http://carnegieendowment.org/1999/07/01/no-more-secrets-policy-implications-of-commercial-remote-sensing-satellites/3lh9.

10. Federation of American Scientists, Intelligence Resource Program, *The Remote Sensing Tutorial*, http://www.fas.org/irp/imint/docs/rst/Front/overview2.html.

11. According to Keith C. Clarke, KH-1 could see a runway, while the improved resolution of KH-4 could see missile silos or cars in a parking lot of the Kremlin. Keith C. Clarke, "America's First Satellite Surveillance," in *The Corona Story*, Project Corona: Clandestine Roots of Modern Earth Science, http://www.geog.ucsb.edu/~kclarke/Corona/story2.htm.

12. *Ibid*.

13. Clarke, "Beyond Surveillance," in *The Corona Story*, http://www.geog.ucsb.edu/~kclarke/Corona/story3.htm#row2.

14. National Reconnaissance Office, "Corona Fact Sheet," http://www.nro.gov/history/csnr/corona/factsheet.html.

15. Details can be found online at http://www.usgs.gov/#/Guides/disp1.

16. NASA, Godard Space Flight Center, "Landsat Data Continuity Mission," updated September 4, 2012, http://landsat.gsfc.nasa.gov/about/ldcm.html.

17. National Aeronautics and Space Administration, "Landsat Then and Now," http://landsat.gsfc.nasa.gov/about/index.html.

18. National Aeronautics and Space Administration, "Landsat 5," http://landsat.gsfc.nasa.gov/about/landsat5.html.

19. Details of the privatization of Landsat are well documented in Florini and Dehquanzada, "No More Secrets?" and in National Aeronautics and Space Administration, "Landsat 5."

20. Florini and Dehquanzada, "No More Secrets?," p. 7 and p. 7 n.20, citing the Land Remote Sensing Policy Act of 1992, section 2(1), Washington, D.C., October 28, 1992.

21. Astrium Geoinformation Services, "Technical Information About the SPOT Satellites," 2012, http://www.astrium-geo.com/na/1240-spot-technical-information; "Pléiades, Very High Resolution Satellite Imagery," 2012, http://www.astrium-geo.com/na/1032-pleiades-very-high-resolution-satellite-imagery ; see also Charlotte Gabriel-Robez, "Seeking Submeter Success," *Earth Imaging Journal* (March-April 2012), http://eijournal.com/2012/seeking-submeter-success.

22. U.S. Geological Survey, "Technical Announcement: SPOT Comes to USGS Archive of Earth Observation Imagery," November 13, 2009, http://www.usgs.gov/newsroom/article.asp?ID=2346.

23. Astrium, SPOTCatalog, http://catalog.spotimage.com/PageSearch.aspx?AspxAutoDetectCookieSupport=1. Prices start at between 1,300 Euros and 8,700 euros per scene.

24. See Astrium, "SPOT Satellite Imagery," http://www.astrium-geo.com/sg/3214-spot-satellite-imagery.

25. William J. Broad, "Giant Leap for Private Industry: Spies in Space," *New York Times*, October 13, 1999, p. A14, http://www.nytimes.com/1999/10/13/us/giant-leap-for-private-industry-spies-in-space.html. There was an Ikonos-1 satellite, but its launch earlier in 1999 failed, and a second satellite, originally named Ikonos-2, was renamed simply Ikonos.

26. *Ibid*.

27. Robert Wright, "Private Eyes," *New York Times Magazine*, September 5, 1999, http://www.nytimes.com/1999/09/05/magazine/private-eyes.html.

28. GeoEye, "Ikonos: Setting the Standard," http://www.geoeye.com/CorpSite/products-and-services/imagery-sources/Default.aspx.

29. GeoEye Commercial Imagery Archive Holdings, http://geofuse.geoeye.com/static/catalog holdings/GeoEye_CatalogHoldings.pdf.

30. Florini and Dehquanzada, "No More Secrets?," p. 10 n.29, citing "Statement by Press Secretary," The White House Office of the Press Secretary, Washington, D.C., March 10, 1994, p. 1.

31. European Space Agency, Earthnet Online, "Ikonos-2," http://earth.esa.int/object/index.cfm?fobjectid=5097.

32. Jennifer LaFleur, "Government, Media Focus on Commercial Satellite Images," *The News Media and the Law* (Summer 2003), p. 37, http://www.rcfp.org/browse-media-law-resources/news-media-law/news-media-and-law-summer-2003/government-media-focus-comm.

33. EOPortal.org, "Quickbird-2," http://www.eoportal.org/directory/pres_QuickBird2.html; DigitalGlobe, Inc., "Products," http://www.digitalglobe.com/products; and Space and Tech, "EarthWatch Granted 'Half-Meter Resolution' License," http://www.spaceandtech.com/digest/sd2001-01/sd2001-01-013.shtml.

34. LaFleur, "Government, Media focus on Commercial Satellite Images," p. 37.

35. Warren Ferster, "NGA Awards Big Satellite Imagery Contracts," *Space News*, August 6, 2010,

http://www.spacenews.com/earth_observation/100806-NGA-awards-imagery-contracts.html.

36. Stephen Wood, "Our Eyes Can't Blink, *The DigitalGlobe Blog*, March 9, 2011, http://www.digital globeblog.com/2011/03/09/our-eyes-can%E2%80%99t-blink.

37. GISuser.com, "GeoEye-1—Visualize Home Plate on a Baseball Diamond from Space!," September 8, 2008, http://www.gisuser.com/content/view/15540/28.

38. GeoEye, "About GeoEye-1," http://launch.geoeye.com/LaunchSite/about/Default.aspx. When I purchased imagery for the *Shades of Green* project, the link above sent me to information about GeoEye-1 which explained that although the imagery is collected at 41 centimeters per pixel, I could purchase it only at 50 centimeters per pixel. Now, in August 2012, the same link leads to information about GeoEye-2, which will be launched in 2013, and the update is this: "GeoEye-2 will have the highest resolution of any commercial imaging system. It will collect images with a ground resolution of 34cm (13.4 inch) in the panchromatic or black-and-white mode. It will collect multispectral or color imagery at 1.36-meter (54 inch) resolution. This advanced resolution will offer our customers unprecedented, precise views for mapping, change detection and image analysis. Although the U.S. government will require GeoEye-2 imagery to be re-sampled from 34cm collection to 50cm products, this will result in better, clearer, sharper 50cm satellite imagery than previously available." In effect this means that the U.S. military will be able to use the imagery at 34 centimeters, or 13.4 inches, while civilians will still only be able to purchase the imagery at 50 centimeters per pixel.

39. Matt O'Connell, "A Leap Forward in Intelligence Gathering: New Developments in Commercial Geospatial Technology Promise Big Payoff," *Defense Systems*, January 20, 2010, http://defensesystems.com/articles/2010/01/27/industry-perspective-geoeye.aspx.

40. Brian X. Chen, "Google's Super Satellite Captures First Image," *Wired*, October 8, 2008, http://www.wired.com/wiredscience/2008/10/geoeye-1-super.

41. ERI, "Overview: What is GIS?," http://www.esri.com/what-is-gis/overview.html#overview_panel.

42. Paul A. Longley, Michael F. Goodchild, David J. Maguire, and David W. Rhind, *Geographic Information Systems and Science: Principles, Techniques, Management, and Applications* (Hoboken, NJ: Wiley, 2005).

43. Steven Johnson, *The Ghost Map: The Story of London's Most Terrifying Epidemic—and How it Changed Science, Cities, and the Modern World* (New York: Riverhead Books, 2006).

44. Robert E. Park and Ernest W. Burgess, *The City: Suggestions for Investigation of Human Behavior in the Urban Environment* (1925; Chicago: University of Chicago Press, 1984).

45. See, for instance, the "GIS Hall of Fame," at the website of the Urban and Regional Information Systems Association (URISA), "the association for GIS professionals," into which McHarg was inducted in the first year (2005): "Arguably, Ian McHarg's 1969 landmark book *Design With Nature* has had a greater influence on the development and application of Geographic Information Systems than any other single event in GIS history." http://www.urisa.org/node/394.

46. Ian L. McHarg, *Design with Nature* (Garden City, NY: The Natural History Press, for the American Museum of Natural History, 1969).

47. Longley et al., *Geographic Information Systems and Science*, p. 16.

48. *Ibid*.

49. *Ibid*., p. 17.

50. Mark Monmonier, *How to Lie with Maps* (Chicago: University of Chicago Press, 1991).

51. Denis Wood, *The Power of Maps* (New York: Guilford Press, 1992).

PROJECTS

1. Jeff Hurn, for Trimble Navigation, *GPS: A Guide to the Next Utility* (Sunnyvale: Trimble Navigation, 1989), pp. 9–10.

2. Paul Virilio, *Desert Screen: War at the Speed of Light*, trans. Michael Degener (London: Athlone, 2002), p. 121.

3. Richard Meier, "Designing the Barcelona Museum of Contemporary Art," in *Richard Meier, Barcelona Museum of Contemporary Art* (New York: Monacelli Press, 1997), p. 16.

4. Claudia Gould, "Interview with Vito Acconci and Steven Holl," in *Vito Acconci and Steven Holl: A Collaborative Building Project, November–December 1993*, reprinted in Joseph Grima et al. (eds.), *Storefront Newsprints 1982–2009* (New York: Storefront Books, 2009), p. 399.

5. GeoLink, "GeoLink Moves Mapping into a Whole New Field — *Yours*" (Billings: GeoResearch Inc., 1993), n.p.

6. Yumiko Ono, "In Japan, They May Never Ask for Directions Again," *Wall Street Journal*, January 7, 1994, p. B1.

7. Hurn, *GPS: A Guide to the Next Utility*, p. 24. Today, the U.S. government refers to GPS as "the world's only global utility." See, for example, the website of the Second Space Operations Squadron at Schriever Air Force Base, Colorado, which operates the GPS system: http://www.schriever.af.mil/GPS.

8. *The Truth*, brochure describing "a field portable remote sensing system" (Arlington: PCI Remote Sensing Corp., 1993), n.p.

9. Fredric Jameson, "The Cultural Logic of Late Capitalism," in *Postmodernism, or The Cultural Logic of Late Capitalism* (Durham: Duke University Press, 1991), p. 54.

10. Intergraph, "Kuwait: Rebuilding a Nation," Intergraph Corp., Huntsville, Alabama, 1991, p. 10.

11. Jean Baudrillard, *The Gulf War Did Not Take Place*, trans. Paul Patton (Bloomington: Indiana University Press, 1995).

12. Virilio, *Desert Screen*, p. 43.

13. Neil Smith, "History and Philosophy of Geography: Real Wars, Theory Wars," *Progress in Human Geography* 16.12 (June 1992), p. 257. Smith, to be fair, was worrying here about the future of geography as a discipline. He was concerned by "the dangerous and self-defeating renunciation of an intellectual (as opposed to technical) agenda that too often accompan[ies] the programmatic advocacy of GIS." "Technology does not *cause* war," he writes, "but the traditional liberal argument that techniques are separate and separable from their uses is equally simplistic." He charges that "liberal advocates" of GIS are "embarrassed into silence by the integration of GIS with military agendas." No silence here, and while the causes of war are beyond the scope of this book, I do advocate the critical uses of these technologies, particularly uses that highlight the qualitative biases inevitably built into them.

14. "Intergraph: Your Partner in Rebuilding Kuwait," advertisement, 1991.

15. "Kuwait City: Image Mapping…the Integration of Remote Sensing, GIS and Digital Cartography," poster DDWA0027A, Intergraph Corp., Huntsville, Alabama, 1991.

16. "How Many Trees Were in Kuwait City?," *Armed Forces Journal International*, June 1991, p. 26.

17. "Kuwait City: Image Mapping…the Integration of Remote Sensing, GIS and Digital Cartography."

18. "A World of Solutions," interview with William E. Salter, *Intervue* (Fall 1991), p. 4.

19. Caryle Murphy, "U.N. Map Makers Draw Kuwaiti-Iraqi Border: Old Documents, New Technology Used," *Washington Post*, May 5, 1992, p. A19.

20. Jeffrey T. Richelson, "U.S. Reconnaissance Satellites Aren't All-Seeing, So Don't Expect Miracles," *Los Angeles Times*, February 17, 1991, p. M5, http://articles.latimes.com/1991-02-17/opinion/op-2094_1_reconnaissance-satellites.

21. "Spacecraft Played Vital Role in Gulf War Victory," *Aviation Week and Space Technology*, April 22, 1991, p. 91.

22. John G. Roos, "SPOT Images Helped Allies Hit Targets in Downtown Baghdad," *Armed Forces Journal International*, May 1991, p. 54; see also Roos, "SPOT's 'Open Skies' Policy Was Early Casualty of Mideast Conflict," *Armed Forces Journal International*, April 1991, p. 32.

23. U.S. News & World Report, *Triumph without Victory: The Unreported History of the Persian Gulf War* (New York: Times Books, 1992).

24. General H. Norman Schwarzkopf, with Peter Petre, *It Doesn't Take a Hero* (New York: Bantam Books, 1992), p. 468.

25. *Armed Forces Journal International*, April 1991, p. 32.

26. Executive Order 12951 of February 22, 1995: Release of Imagery Acquired by Space-Based National Intelligence Reconnaissance Systems, *Federal Register* 60.39 (February 28, 1995), pp. 10789–90, http://www.gpo.gov/fdsys/pkg/FR-1995-02-28/pdf/95-5050.pdf.

27. Oliver Morton, "Private Spy," *Wired* 5.8 (August 1997), pp. 114–99 and 149–52; http://www.wired.com/wired/archive/5.08/spy.html.

28. Keith C. Clarke, "Beyond Surveillance," in *The Corona Story*, Project Corona: Clandestine Roots of Modern Earth Science, http://www.geog.ucsb.edu/~kclarke/Corona/story3.htm.

29. Jim Graham, Lockheed-Martin Space Corp., "Corona Program Profile," May 1995; distributed in *Space News Digest* 1.298 (June 16, 1995), http://www.islandone.org/SpaceDigest/SpaceDigestArchive/SortingInProgress/SpaceNewsDigest.v01/v1n0298.

30. Kevin Ruffner (ed.), *Corona: America's First Satellite Program* (Washington, D.C.: Central Intelligence Agency, Center for the Study of Intelligence, 1995).

31. See Gillian Cook, "Khayelitsha: New Settlement forms in the Cape Peninsula," in David M. Smith (ed.), *The Apartheid City and Beyond* (London: Routledge, 1992), pp. 125–35. The South African Census of 2001 counts the population at 329,000, http://www.capetown.gov.za/en/stats/2001census/Documents/Khayelitsha.htm. But this number is famously underestimated. The Western Cape Population Unit's "Population Register Update: Khayelitsha 2005," published in 2006, explains the undercounting in the 2001 census and corrects it with results from a 2004 aerial survey. The report can be found at http://www.westerncape.gov.za/other/2007/10/kprufinal_2005_october_2007_publish_date.pdf.

32. See Denis Cosgrove, *Apollo's Eye: A Cartographic Genealogy of the Earth in the Western Imagination* (Baltimore: Johns Hopkins University Press, 2001), pp. 257–62, for a discussion of the new ways of seeing the Earth as catalyzed by the Apollo 8 and Apollo 17 photographs.

33. "Flashbacks," *Life Magazine*, October 1999, p. 66.

34. "All these pictures go on the Web site, so this will appear on the Web site," said the Pentagon spokesman. Office of the Assistant Secretary of Defense (Public Affairs), News Transcript, "DoD News Briefing, Wednesday, June 9, 1999 — 2:20 p.m.," http://www.defense.gov/transcripts/transcript.aspx?transcriptid=483.

35. Charles Lane and Thom Shanker, "Bosnia: What the CIA Didn't Tell Us," *New York Review of Books*, May 9, 1996, p. 10, http://www.nybooks.com/articles/archives/1996/may/09/bosnia-what-the-cia-didnt-tell-us.

36. Office of the Assistant Secretary of Defense (Public Affairs), News Transcript, "DoD News

Briefing, Saturday, April 10, 1999—2:05 p.m.," http://www.defense.gov/transcripts/tran
script.aspx?transcriptid=583.

37. "On the way to the bus station…a police officer said to me: 'The war started in Drenica,
    and we are going to end it here.'" Glogovac resident, quoted by Human Rights Watch,
    "'Ethnic Cleansing' in the Glogovac Municipality," July 1999, http://www.hrw.org/legacy/
    reports/1999/glogovac.

38. SPOT has since been acquired by Astrium. The SPOT catalogue is online at: http://catalog.
    spotimage.com/PageSearch.aspx?AspxAutoDetectCookieSupport=1.

39. Thanks to Branden Joseph for pointing out image 05.20.99/083-264 when I presented this
    work as part of a lecture called "Random Access Memory" at a conference at Princeton Uni-
    versity in 2000. He asked me to return to what he called the "white on white" image in my
    slides. His comment got me thinking about a future project titled *Monochrome Landscapes*.

40. Cloud cover preoccupied NATO and the Pentagon during the Kosovo campaign in terms of
    both the bombing campaign and the effectiveness of overhead imaging systems, which is also
    to say the public presentation of the battle. See Paul Watson, "Dispatch from Kosovo: Break
    in Clouds Can Give Allies Clear View of Targets," *Los Angeles Times*, April 3, 1999, p. A9. Later,
    in an interview, Kenneth Bacon, the Pentagon spokesman, said that "heavy cloud cover over
    Yugoslavia meant little aerial photography was available to show in the first weeks." Brad-
    ley Graham, "Pentagon's News Filter May Obscure Air War Effect," *Washington Post*, May
    24, 1999, p. A20. After the war, Bacon suggested that one of the war's lessons was the need
    for "drones, particularly ones that can somehow penetrate through foliage and through bad
    cloud cover. So we all were hampered somewhat by the bad weather and by the intense for-
    estation, or foliage, in Kosovo and Yugoslavia. So we're looking for ways that we can pene-
    trate that more effectively in the future." Office of the Assistant Secretary of Defense (Public
    Affairs), News Transcript, "USIA Foreign Press Center Briefing—Kenneth H. Bacon," October
    21, 1999, http://www.defense.gov/transcripts/transcript.aspx?transcriptid=336.

41. Cable News Network, "America Strikes Back: Pakistan Warns Northern Alliance; Second Wave
    of Attacks Under Way," transcript, CNN Live Event/Special 14:20, Monday October 8, 2001,
    Transcript #100823CN.V54 (Lexis/Nexis).

42. Gilles Peress et al., *Here Is New York: A Democracy of Photographs* (New York: Scalo Publish-
    ers, 2002).

43. Ed Vulliamy, "A Mass Grave or Prime Real Estate?," *Guardian*, March 9, 2002, http://www.
    guardian.co.uk/world/2002/mar/10/terrorism.september113.

44. A Lexis/Nexis search indicates that the image was published on September 14 by newspa-
    pers as diverse as the *South China Morning Post* in Hong Kong and London's *Daily Mail*. See
    also Barnaby J. Feder, "Bird's-Eye Views, Made to Order," *New York Times*, October 11, 2001,
    http://www.nytimes.com/2001/10/11/technology/bird-s-eye-views-made-to-order.html:
    "Among the indelible images from the terrorist attacks last month were commercial satellite
    photographs showing smoke and dust of volcanic proportions stretching for miles from the
    ruins of the World Trade Center."

45. Miroslav Prstojevic et al., *Survival Guide Sarajevo* (Sarajevo: FAMA; New York: Workman
    Publishing, 1993); Nihad Kresevljakovic et al., *Survival Map 1992–1996* (Sarajevo: FAMA, 1996).
    FAMA's innovative and intelligent work during the war and the siege, including both these
    documents, is now online at http://www.famacollection.org/eng/fama-collection/fama-
    original-projects/index.html.

46. We used some sentences from Christy Ferer, widow of Port Authority executive director Neil Levin, who was killed on September 11, as a sort of epigraph on the map. She drew a distinction between "gawkers" and "tourists," on the one hand, and those who "have come to ground zero to pay respects and to deal with the psychic blow of what happened here." We weren't sure that distinction was always so clear, and so we quoted the last words of her *New York Times* op-ed. "It is a burial ground. It is a cemetery, where the men and women we loved are buried. Where they rest is now hallowed ground." Christy Ferer, "Unforgotten Soldiers," *New York Times*, October 25, 2001, p. A21, http://www.nytimes.com/2001/10/25/opinion/unforgotten-soldiers.html.

47. On the International Criminal Tribunal for the former Yugoslavia, see Laura Kurgan, "Residues," in *Alphabet City 7: Social Insecurity* (September 2000), pp. 112–30.

48. Ellsworth Kelly, *Four Panels* (1970–71). Screenprint on Special Arjomari paper, 36-3/4 x 62 inches (93.3 x 157.5 cm). There are many Ellsworth Kelly monochrome paintings and prints, but this is the one I had in mind.

49. This coincidence was noted by David Firestone, "Drilling in Alaska, a Priority for Bush, Fails in the Senate," *New York Times*, March 20, 2003, p. A1, http://www.nytimes.com/2003/03/20/us/drilling-in-alaska-a-priority-for-bush-fails-in-the-senate.html; Bob Herbert, "Ready for the Peace?," *New York Times*, March 20, 2003, p. A31, http://www.nytimes.com/2003/03/20/opinion/ready-for-the-peace.html.

50. U.S. Geological Survey, Fact Sheet 0028-01, "Arctic National Wildlife Refuge, 1002 Area, Petroleum Assessment, 1998, Including Economic Analysis," http://pubs.usgs.gov/fs/fs-0028-01/fs-0028-01.htm.

51. "International Meridian Conference," in *Annual Report of the Board of Regents of the Smithsonian Institution, Showing the Operation, Expenditures, and Condition of the Institution for the Year 1884* (Washington, D.C.: Government Printing Office, 1885), p. 186. See Peter Galison's account of the conference and the "struggle for symbolic centrality" in *Einstein's Clocks, Poincaré's Maps* (New York: W. W. Norton, 2003), pp. 144–55.

52. See Mark Monmonier, *Drawing the Line: Tales of Maps and Cartocontroversy* (New York: Henry Holt, 1995), ch. 1.

53. Natural Earth is a collaborative effort by volunteers to release map data for free. See "Natural Earth Version 1.3 Release Notes," http://www.naturalearthdata.com/blog/natural-earth-version-1-3-release-notes. I first learned about this warning on http://roomthily.tumblr.com/post/3041306314/null-island.

54. I was inspired to imagine the possibility of doing civilian satellite investigations by the before-and-after images of Grozny in the *New York Times* in March 2000.

55. "The 1994 Forest Code calls for all commercial logging to be regulated under designated forest concessions. Before they can be legally logged, areas slated for timber production are allocated to timber operators under a defined selection process." Benoit Mertens et al., *Interactive Forestry Atlas of Cameroon, Verion 2.0* (2007), p. 3, a Global Forest/World Resources Institute watch report, http://www.wri.org/publication/interactive-forestry-atlas-cameroon-version-2-0.

56. *Ibid.*, p. 5.

57. "Each forest concession with a MINFOF–approved management plan is divided into 30 logging parcels (i.e., AACs), which are integral to the 30-year logging rotation that is at the heart of the sustainable forest management process." *Ibid.*, p. 9.

58. See *ibid*.

59. I wrote about this in "Trying Not to Avoid Propositions Altogether," *Assemblage* 41 (April 2000), p. 37.

60. Gary Wolf, "Exploring the Unmaterial World," *Wired* 8.6, (June 2000), pp. 302–19, http://www.wired.com/wired/archive/8.06/koolhaas_pr.html.

61. Rem Koolhaas, "Earning Trust," lecture at a conference on Superhumanism in London in 2001, quoted in Doerte Kuhlmann, "Big Bright Green Pleasure Machine," in Karin Jaschke and Silke Ötsch (eds.). *Stripping Las Vegas: A Contextual Review of Casino Resort Architecture* (Weimar: University of Weimar Press, 2003), p. 180. A decade later, Koolhaas expressed some doubts about this, telling an interviewer from London's *Independent*: "We're very glad that what I call the 'YES' regime—which means the Yen, the Euro and the US dollar—which began in the 1980s and that dictated every value in every country, has finally come to an end. And I think it's a very good thing that the state is becoming responsible again after a long time of deregulation." Susie Rushton, "The shape of things to come: Rem Koolhaas's striking designs," *The Independent*, June 21, 2010. http://www.independent.co.uk/arts-entertainment/architecture/the-shape-of-things-to-come-rem-koolhaass-striking-designs-2005994.html.

62. Francis X. Clines, "Ex-Inmates Urge Return To Areas of Crime to Help," *New York Times*, December 23, 1992, p. A1, http://www.nytimes.com/1992/12/23/nyregion/ex-inmates-urge-return-to-areas-of-crime-to-help.html.

63. Lola Odubekun, *Vera Institute Atlas of Crime and Criminal Justice in New York City* (New York: Vera Institute of Justice, 1993), pp. 42 and 38.

64. *Ibid.*, p. 43.

65. Jennifer Gonnerman, "Million-Dollar Blocks: The Neighborhood Costs of America's Prison Boom," *Village Voice*, November 9, 2004, pp. 17–23, http://www.villagevoice.com/2004-11-09/news/million-dollar-blocks.

66. Jeremy Travis, *But They All Come Back; Facing the Challenges of Prisoner Reentry* (Washington, D.C.: The Urban Institute Press, 2005), p. 5.

67. James Austin et al., *Unlocking America: Why and How to Reduce America's Prison Population* (Washington, D.C.: The JFA Institute, November 2007), p. 1, http://www.jfa-associates.com/publications/srs/UnlockingAmerica.pdf.

68. National Institute of Justice, *What is Crime Mapping?: Briefing Book* (2005), p. 1, http://www.cops.usdoj.gov/html/cd_rom/tech_docs/pubs/WhatIsCrimeMappingBriefingBook.pdf.

69. *Ibid.*, pp. 1 and 4.

70. The Bureau of Justice Statistics updates its prison statistics each year. Go to http://bjs.ojp.usdoj.gov and click on "Corrections." You will find a link to "Total Correctional Population."

Design, typesetting, and production by Julie Fry

Printed and bound by Friesens, Altona, Manitoba